# The Jukebox

# Nature, Culture and Literature
# 07

# The Jukebox in the Garden

## Ecocriticism and American Popular Music Since 1960

David Ingram

Rodopi

Amsterdam - New York, NY 2010

Cover image: www.morguefile.com

Cover design: Pier Post

The paper on which this book is printed meets the requirements of "ISO 9706:1994, Information and documentation - Paper for documents - Requirements for permanence".

ISBN: 978-90-420-3209-5
E-Book ISBN: 978-90-420-3210-1

Printed in the Netherlands

*For my parents, Roy and Eileen*

# Table of Contents

I would like to thank the following for helping me in various ways along the way: Scott Slovic, Geoff Ward, Terry Gifford and Axel Goodbody for their support, and the two anonymous referees at Rodopi for their generous and insightful comments. Early drafts of the book were first aired at conferences organised by the Association for the Study of Literature and Environment at Boston University (2003), University of Oregon, Eugene (2005) and the University of Edinburgh (2008), the British Association for American Studies at the University of Oxford (2002), the University of Aberystwyth (2003) and Manchester Metropolitan University (2004) and at the University of Queensland (2002). My thanks go to the organisers of these excellent conferences. I acknowledge the support of the Arts and Humanities Research Council for a Research Leave grant in 2003-4.

An early version of part of Chapter 6 was published as '*My Dirty Stream*: Pete Seeger, American Folk Music, and Environmental Protest' in *Popular Music and Society*. 31.1 (2008): 21-36.

An early version of part of Chapter 7 was published as '*Go to the forest and move*: 1960s American rock music as electronic pastoral' in *The Forty-Ninth Parallel: An Interdisciplinary Journal of American Studies*. 2 (Spring 2007): 1-16.

# Introduction

The rise of the modern environmental movement in the 1960s saw the notion of ecological crisis enter popular debates on the natural and built environments. This book traces the various ways in which American popular musicians reacted to such developments. The historical context is well known. In 1962, biologist Rachel Carson raised fears of environmental catastrophe by drawing attention to the use of pesticides in industrial agriculture. Fears of overpopulation were highlighted with the publication of Paul Ehrlich's *The Population Bomb* in 1968. A series of environmental accidents, including the Santa Barbara oil spill in January 1969 and the fire on the Cuyahoga River, Ohio in June of the same year, added to public concerns. For the New Left, opposition to the Vietnam War turned into a wider belief that capitalism was waging war against the Earth itself, while some members of the hippie counterculture responded to the social and political crisis by setting up rural communes. As the 1970s saw this back-to-nature movement transmute into New Age environmentalism, the first Earth Day in April 1970 signalled the beginning of the institutionalistion of the modern environmental movement (Rossinow 1998: 276). The ensuing decades have seen a catalogue of environmental issues brought to popular attention, from hunger, nuclear power, species extinction, the destruction of the rainforests and ozone depletion to climate change.

American popular musicians have responded to these concerns in a variety of ways. Environmental issues have become the subject matter for popular music, and songwriters and composers in a wide range of styles, from folk singer Pete Seeger to jazz saxophonist Paul Winter, have lamented, and protested against, what they see as the degradation of the Earth. *The Jukebox in the Garden* explores how environmental themes have been represented in popular song. It also investigates the growing link between music and ecophilosophical thought, according to which music, amongst all the arts, has a special affinity with ecological ideas. Unsurprisingly, much explicit theorising about music and ecology has come from advocates of 'art' music, with composer John Cage most influential in such debates. Cage rejected 'program' music, in which composers seek to imitate either the literal sounds of the natural world or the subjective effects that particular landscapes have on them. Instead, music was for him an attempt to imitate what he understood to be the inner workings of the natural world, rather than its external appearances. He quoted the Indian art critic Ananda K. Coomaraswamy, who wrote that the traditional role of the artist in Indian art was 'to imitate Nature in her manner of operation'. Coomaraswamy had taken this phrase from the medieval theologian St. Thomas Aquinas: imitation, in Aquinas' neo-Platonist sense, meant not copying the external appearances or concrete details of the natural world, but its inner, essential forms (Cage 1961: 194; 1968: 31). Cage drew on such philosophical ideas about nature and ecology as a creative model for both composition and performance. Asserting in the early 1970s that 'Music IS Ecology', he developed revolutionary notions of musical form influenced by Taoist and Zen Buddhist conceptions of the natural world (1981: 229). In effect, instead of writing and playing music *about* ecology, Cage wanted to make ecological music; his music aspired to be a form of ecophilosophical speculation in itself (Ingram 2007a).

The popular musicians discussed in this book tend to be less explicitly theoretical than art composers like Cage. Popular music, as musicologist Theodore Gracyk points out, tends to arise 'from the materials, not from theory' (1996: 119). Nevertheless, as we will see, ecophilosophical speculation is also emerging in American popular music, particularly under the influence of both deep ecology and New Age thinking about the relationship between human beings and the rest of the natural world. Deep ecology has influenced composer, clarinettist and philosopher David Rothenberg, who has written most extensively about music and ecological thought. As we will see in Chapter 12 on Jazz, improvisation is for Rothenberg the key to his artistic strategy of 'looking for nature right in the music' (2001: 7). Musicologist Charles Keil's *Born to Groove* (2006) is also influenced by deep ecology, and provides another sustained exploration of the idea that playing music can be a route to ecological awareness. More unashamedly New Age thinkers on music and ecology include the British composer and music therapist June Boyce-Tillman and the American 'neo-pagan' Jesse Wolf Hardin, otherwise known as Lone Wolf Circles. This book is, in part, an exploration and critique of such speculations on the role that music can play in raising ecological awareness.

In some respects, such claims are a revival of ideas about the efficacy of music which date back in Western Europe to Pythagorean notions of the 'harmony of the spheres' in ancient Greece. The Pythagoreans, writes musicologist Enrico Fubini, 'looked upon the soul as a harmony. Following on from this, music was credited with a special influence over the soul in view of music's kinship with the universal harmony in whose image the soul, too, was thought to be constituted. What is more, music could bring about a re-establishment of harmony within men's souls when they were in a troubled state' (1987: 25). A more direct link with the theoretical models of music discussed in this book is with nineteenth-century German idealist philosophy, in which, as Christopher Norris writes, music was considered the highest form of art 'on account of its unique expressive power, its capacity to fuse the phenomenal sound-world of sensuous experience with a sense of some ultimate significance beyond the grasp of mere reason' (1989: 320). The ideas of Arthur Schopenhauer, in particular, can be traced in Ernst Bloch's speculations on tonal music as a form of utopian striving, as Chapter 1 will show.

It is with such twentieth- and twenty-first century thinkers that this book is mainly concerned. Between the 1920s and the 1960s, humanist Marxists Bloch and Theodor Adorno created a significant body of critical work, both in Germany and the United States, on the philosophical relationship between music, politics and the natural world. Bringing together ethics and aesthetics, they broadly shared a belief in music as a utopian art whose forms symbolically prefigured the socialist society they hoped for in the future. This society would be based on a natural world tamed for human purposes by modern, industrial technology. Aesthetically, neither critic valued popular music: Bloch tended to favour the canonical works of the Western classical tradition, while Adorno was a leading advocate of twentieth-century modernism.

By the early 1970s, some radical environmentalists began to respond to the newly perceived ecological crisis by criticising humanist Marxism's unquestioned faith in the benefits of industrial progress and economic growth. This critique eventually began to inform debates in music aesthetics. The 'new musicology', which emerged from Susan McClary and others in the late 1980s, formulated a postmodern critique of the modernist aesthetics advocated by Adorno. The new field of ecopsychology has been particularly influential in the development of the new eco-aesthetics; the art critic Suzi Gablik's *The Reenchantment of Art* (1991) consciously promoted an awareness of holistic relationality in visual art, in keeping with what many radical environmentalists assume to be a vital prerequisite for ecological health and sustainability. It is worth establishing the philosophical bases of these developments in detail.

Gablik's book broke new ground in both challenging the aesthetic orthodoxies of political modernism and in developing an explicitly ecological theory of art. She argues that the values of modernism, particularly its emphasis on confrontation and alienation, have not only been co-opted by the capitalist commodification of visual art in the late twentieth century, but have also outlived their effectiveness in a contemporary age of ecological crisis. The 'psychic and social structures in which we live have become too profoundly antiecological, unhealthy and destructive', she writes. 'There is a need for new forms emphasising our essential interconnectedness rather than our separateness, forms evoking the feeling of belonging to a larger whole rather than expressing the isolated, alienated self' (Gablik 1991: 62).

For Gablik, the new ecological sensibility is embodied artistically in the work of land artist Lynne Hull, who makes small, intimate outdoor sculptures which act as wildlife habitats, and by the Canadian composer and 'acoustic ecologist' R. Murray Schafer, who performs pieces such as his opera *Princess of the Stars* (1985) outdoors, rather than in concert halls, in order to reconnect his listeners with living environments (1991: 86-9). These artists have replaced the stance of hostility and aggression that modernist art takes towards bourgeois society with more affirmative notions of social responsibility and mutual interactivity.

Underpinning the new eco-aesthetics is the concept of the 'ecological self'. The argument is as follows. Liberalism assumes a notion of self that is atomistic (that is, separate from the world around it), individualistic and self-interested; what the seventeenth-century British philosopher John Locke called 'possessive individualism' (McPherson 1962). One of the beliefs that unites many different branches of radical environmentalism, including deep ecology, ecofeminism, ecosocialism and ecoanarchism, is that this liberal self has had dire consequences for the health of the environment, because it has encouraged in modern Western societies both an anthropocentric disregard for other organisms, and an acquisitive materialism that risks exhausting the Earth's natural resources through overdevelopment. In opposition to the liberal humanist self, radical ecophilosophers have proposed an 'ecological' self that is relational rather than atomistic, and thus more likely to consider the interests of other organisms, and to have a less destructive impact on the environment than the liberal self. Influential theorists of the ecological self include deep ecologist Arne Naess and ecofeminist Freya Matthews (Naess 1989; Matthews 1999).

The main problem with the theory is the assumption that the concept of holism provides a scientific basis for what are essentially political speculations. Yet holism is itself a contested concept with science. From its origins in the middle of the nineteenth century, the science of ecology has tended to advocate a holistic understanding of ecosystems; that is, the idea that individual organisms in a particular habitat are connected to a whole that is greater than the sum of its parts. In *The Closing Circle*, the biologist Barry Commoner famously expressed this concept in his First Law of Ecology, that 'everything is connected to everything else' (1971: 33). In the late 1960s, however, scientists influenced by the newly emerging chaos theory began to challenge the dominance of the holistic paradigm. According to the new thinking, real ecosystems may be too complex to act as coherent wholes, and are better characterisd by patchiness rather than overall coherence. 'Significant numbers of the living creatures found in any given habitat', ecocritic Dana Phillips explains, 'are likely not to be integral participants in whatever large-scale phenomena may be occurring in their habitat day after day' (2003: 66). Despite these reservations, however, ecophilosophical speculation based on the supposedly scientific notion of the 'ecological self' has prospered, particularly since its adoption by ecopsychology in the 1990s (Roszak 2001). It informs the postmodernist aesthetics of music which we will explore in this book.

Taken as a whole, Part One of *The Jukebox in the Garden*, 'Theories of ecocriticism and popular music', examines these ecophilosophical claims for music in detail, and establishes the key concepts and theories that inform the rest of the book. Chapter 1, 'Popular music and environmental ethics' explores the claim that music is a form of utopian expression that prefigures a better society in the future, including a healed relationship between human beings and the natural world. It focuses mainly on the theories of Ernst Bloch and the later radical environmentalist critique of humanist Marxism. The critical writings of musicologist Susan McClary, composer and music therapist June Boyce-Tillman, and French post-structuralist philosophers Gilles Deleuze and Félix Guattari are also discussed as contributions to the development of philosophical speculation on the relationship between music and ecological thinking. Chapter 2 goes on to discuss the particular role of the 'popular' in popular music, moving beyond Theodor Adorno's negative view of popular music through the work of Richard Dyer and Jason Toynbee on the utopian promise of popular entertainment. Chapter 3 examines the use of the concept of 'nature' in music criticism, particularly the claim that authentic music is produced by performers who are close to the natural world. The chapter ends with a discussion of the vital role played by the pastoral mode in representations of the natural world in North American popular music, by drawing on the ecocriticism of Leo Marx and Lawrence Buell. Chapter 4, 'Eco-listening', looks at the claim made by some eco-philosophers and musicians that the activity of listening itself has a special role to play in the formation of ecological awareness. Theories of rhythmical 'entrainment' and 'immersive listening' have been given an ecological focus by deep ecological and New Age thinkers such as Charles Keil and William Irwin Thompson.

Having established a conceptual framework for the book, Part Two surveys a wide range of popular musical styles to examine the different ways in which they have mediated American relationships between nature, technology and environmental politics. Each chapter has its own central theme relevant to the musical style in question. Chapter 5 examines the strengths and limitations of the proletarian realist aesthetic on which topical folk music has been based, from the mid-sixties work of Pete Seeger to the recent contributions of David Rovics and others. It also looks at the parallel folk tradition which guitarist John Fahey called the 'American Primitive', recently revived by Devendra Banhart and Joanna Newsom, which relates to the natural world as a place of mystical enchantment, rather than a cause to be explicitly defended. Chapter 6 examines claims about the natural authenticity of both African-American blues and country music, and analyses the recurrent tropes of the farmer and cowboy that are central to the appeal of the latter. Chapter 7, on rock music and R'n'B in the 1960s, exposes contradictions in the sixties rock counterculture's 'electronic pastoral', which simultaneously embraced both a desire to return to nature and the high tech futurism of the space age. Chapter 8 examines the ironic use of the pastoral mode in the country rock movement that emerged in the late 1960s. The environmentalist implications of rock music, as the values of the 1960s counterculture were challenged by punk in the mid-1970s and beyond, are the subject of Chapter 9. The final chapters of the book survey issues related to three broad areas of musical production. Chapter 10 looks at how globalisation has produced a 'One World' ideology of ecological interconnectedness in world music; Chapter 11 discusses electronic music from ambient electronica to the rave scene of the 1990s, to investigate how the use of the digital computer in music-making questions Romantic ideas that link creativity with organicism; Chapter 12 investigates how the improvised elements in jazz have provided a model for a benign relationship between human beings and the natural world. David Rothenberg's hope that playing his clarinet to birds and whales will lead to interspecies communication is one of the more speculative manifestations of a belief in the power of music to both enact and foster greater ecological awareness in its audiences.

Ecocriticism, like the radical literary theory from which it emerged in the late 1980s, has tended to take a moralist stance towards works of art; that is, it views the ethical merit (or flaw) in a work of art as also an aesthetic merit (or flaw). A central concern has therefore been what might constitute an ecologically progressive work of art. For Lawrence Buell, such a work will be one in which, 'Human accountability to the environment is part of the text's ethical orientation' (1995: 7-8). But it is counter-productive to be overly prescriptive when it comes to art. As Henri Lefèbvre observed, prescriptions are the counterpart of prohibitions (1991: 210). Taken to its extreme, prescriptions and blueprints lead to dogmatism and authoritarianism.

Aesthetic philosopher Berys Gaut shows how to escape from such strictures, by arguing for an aesthetically pluralist form of moralism. He uses Leni Riefenstahl's pro-Nazi film *The Triumph of the Will* (1935) to illustrate his case. The moralist, he writes, should be 'a pluralist about aesthetic values; and then he or she can hold that the film is aesthetically flawed insofar as it is immoral, but that it has many aesthetic merits, such as its unity and complexity, which overall make it a good work of art' (Gaut 2001: 344). This aesthetic method has the advantage of being nuanced, generous and non-dogmatic; it is open to a variety of artistic approaches while avoiding the overly prescriptive. It is with this critical framework in mind that I have approached the popular music featured in this book.

*The Jukebox in the Garden* places North American popular music made during the era of modern environmentalism within its wider social, historical and political contexts, augmenting the lyric analysis and musical description that constitute the main approach taken in the book. Ultimately, of course, musicians communicate through their music, but their attempts to describe their work in words can also shape the way it is listened to and appreciated. Moreover, although a listener's reaction to a piece of music transcends the intentions of its creators, knowing the latter's stated intentions can nevertheless illuminate the work and extend both pleasure and understanding. It is in this spirit that the explorations in this book are intended.

The stylistic typology I have used to organise the chapters in Part 2 is necessarily provisional and fuzzy. Overlaps abound. For example, rock music and world music often include an element of improvised performance, and are therefore relevant to the discussions in Chapter 12. Jon Hassell is a good example of a musician who crosses stylistic boundaries, and could have been placed in the chapters on jazz or electronica, as well as the one on world music in which I have chosen to place his work. Similarly, David Axelrod's *Earth Rot* (1970) is a mixture of jazz, rock, light classical and show tunes that belies categorisation. Of course, the very notion of the 'popular' is contested, and raises questions about both musical form and contexts of production and reception. It may be that David Rothenberg and Jon Hassell do not play 'popular' music according to many definitions, but their stylistic overlap with jazz and electronica allows for their inclusion in a book that is not intended to be too obsessed with debates about categorisation.

For the purposes of this book, then, 'pop' includes jazz and electronica, and is not a separate category from rock, R'n'B or hip hop. By being so inclusive, my aim has been to reflect the diversity of popular music produced in the United States since about 1960. The focus is on 'white' and African-American music in particular: Native American and Hispanic popular music deserve in-depth studies of their own, and are beyond the scope of the book. Of course, many pop records celebrate and endorse consumerist values, both explicitly and implicitly, and are therefore ripe for ecocritical analysis; the critical emphasis here, though, is on more or less explicit representations of either ecology or the natural world in popular music.

The music covered in *The Jukebox in the Garden* reveals a musical culture characterised by stylistic hybridity and a healthy defiance of critical categories. Indeed, aesthetic debates about music, in which the proletarian realism of protest folk music meets the intuitive hedonism of rock, are a unifying theme of the book. Although some musicians impute eco-utopian potential to music in general, in practice most show aesthetic preferences for particular types of music which they see as ecologically healthy. The popular musicians featured in this book have embraced the pleasures offered by a range of musics: New Age, folk, world, jazz, ambient, rock and electronica. As we will see, all of these musics have their advocates as means of fostering an environmental ethics of care in their audiences.

Although our perception of music as a physical phenomenon depends on natural principles of acoustics and physiology shared by all listeners, the meanings that human beings project onto it are relative, both to particular individuals and to the wider cultures and subcultures in which they are located. Musicologists Shepherd and Wicke describe two levels of signification in music: the 'syntactic' is concerned with how individual notes relate to each other, whereas the 'semantic' concerns the connotative impressions created by music (1997: 103). It is with this second area that this study is mainly concerned. Its emphasis is therefore less on close musicological analysis than on the wider historical, social and cultural contexts in which the production and consumption of music takes place, themselves shaped by factors such as class, race, gender and geographical location.

The musicologist Michael Chanan writes that music is 'so diverse and kaleidoscopic' that there is 'no individual theoretical framework and no single discipline which can adequately comprehend it' (1994: 9-10). With this in mind, *The Jukebox in the Garden* analyses a wide range of ecophilosophical discourses produced by composers, musicians, philosophers, musicologists, historians, critics and fans. The focus on the United States, like the post-1960 time-frame, is a necessary way of limiting the field, rather than a statement about the cultural importance, or otherwise, of the nation itself. The book does not explore in detail a political economy of popular music, or the ecological effects of the production and consumption of music, a topic I have attempted to address elsewhere (Ingram 2007b).

Writing about music is necessarily impressionistic. I have tried to describe the intellectual and emotional effects that individual pieces of music have had on me, while avoiding the hyperbole often associated with popular music journalism. The field covered is wide ranging, but inevitably partial. Some musicians and styles are featured more extensively than others. I have selected individual pieces of music to represent what I see as noteworthy creative strategies, either because they are innovative or because they are typical of a wider field. Inevitably, such a selection process is personal, contingent and open to challenge. My hope is that readers will use the book to explore their own pathways through this vast and largely unexplored topic.

# Part 1

# Theories of ecocriticism and popular music

# Chapter One

# Popular music and environmental ethics

Philosophical speculation on the relationship between music aesthetics and an ethics of the environment has taken several forms. This chapter explores the three main theoretical approaches, which may be broadly characterised as the humanist Marxism of Ernst Bloch and Theodor Adorno, the postmodernist musicology of Susan McClary and June Boyce-Tillman, and the post-structuralism of Gilles Deleuze and Félix Guattari. Each of these thinkers takes up a different ethical position regarding a key concept that informs this book: the utopian potential of music to prefigure a good society, including a more benign relationship between human beings and the natural world. They also hold differing aesthetic positions in relation to the notion of the 'popular' in music. After investigating the strengths and weaknesses of the three theoretical models, the chapter ends by proposing an alternative philosophical model for ecocriticism based on critical realism. This model will guide the critical readings of popular music that make up Part Two.

## 1.1 Humanist Marxism: Ernst Bloch

In his major work *The Principle of Hope*, which he completed in 1958, Ernst Bloch expressed his faith in the Marxist assumption that the key to social and political progress lies in extending human control over the natural world through modern, industrial technology. The arts, especially music, were for Bloch a utopian prefiguration of the future good society that would be fulfilled in a socialist state.

Bloch's claims for music can be traced back to the nineteenth-century German philosopher Arthur Schopenhauer. He was influenced by the Vedantic doctrine of 'maya', according to which our everyday experience of the world as being made up of separate, discrete objects (which he called the *principium individuationis* or 'principle of individuation') is an illusion. What is real, he claimed, is the numinous realm of the Will, or will-to-life, which drives all matter in the universe. Schopenhauer believed that music, because of its non-verbal nature, could reveal this transcendental reality better than any other art form. The composer, he wrote, 'reveals the innermost nature of the world, and expresses the profoundest wisdom in a language that his reasoning faculty does not understand' (Schopenhauer 1969: 260). For Schopenhauer, the sense of striving created by tonal music, as the melody moves away from, and finally back towards, the keynote, expresses a utopian yearning for a final destination or goal that is endlessly deferred. 'The inexpressible depth of all music', he wrote, 'by virtue of which it floats past us as a paradise quite familiar and yet eternally remote, and is so easy to understand and yet so inexplicable, is due to the fact that it reproduces all the emotions of our innermost being, but entirely without reality and remote from its pain' (264).

Bloch reinterpreted Schopenhauer's conception of music through a Marxist critique of capitalism. Unusually for a Marxist, he based his conception of nature on the distinction, made by the twelfth-century Andalusian-Arab philosopher Averroës (Ibn Ruschd), and later developed by Baruch Spinoza, between 'natura naturata' and 'natura naturans': the former is the passive, inert matter which makes up everyday reality, whereas the latter is an active, creative force that resides within matter. This vitalist theory was adopted by medieval alchemists such as Paracelsus, but was eventually marginalised in the seventeenth century by the new materialist physics of Galileo and Newton (Merchant 1982: xix). Yet Bloch was keen to reconcile the concept of 'natura naturans' with the findings of modern science, and therefore to distance it from what he called the 'pantheistic bogeyman': for him, it was a secular, materialist notion that referred to the objectively knowable, physical processes of the natural world as known to contemporary science (Bloch 1995: 671). He believed that science, technology and the arts were all means of understanding and connecting with these fundamental natural processes, thereby enabling human beings to use 'natura naturans' as a resource for their own well-being. Socialism represented the best means of achieving what Marx called this longed-for 'realm of freedom'. A key text for Bloch was the early Marx of *The Economic and Philosophical Manuscripts of 1844* on the necessity of human labour to improve the world (Bloch 1995: 112). In Bloch's version of Marx, the goal of human labour remained 'the naturalisation of man, humanisation of nature which is inherent in developing matter' (209).

Bloch also traced the project of humanising nature through technology to the seventeenth-century natural scientist Francis Bacon, who wrote of his utopian desire to establish a *regnum hominis*, or kingdom of man, on Earth, by fulfilling his maxim, 'Natura parendo vincitur': Nature is conquered by obedience (1995: 657). According to Bloch, Bacon's program only led to the over-exploitation of nature because of the inequalities of bourgeois social relations within which science and technology have been practiced. In a future revolutionary socialist era, human beings would be properly allied with 'natura naturans', so that their control of nature would be not only more effective than previously, but also benign and non-exploitative. As he put it, the capitalist concept of technology 'exhibits more domination than friendship, more of the slave-driver and the East India Company than the bosom of a friend' (1995: 670). In contrast, 'Marxism of technology', when properly developed, will see 'the end of the naive application of the standpoint of the exploiter and animal tamer to nature' (1995: 695).

*The Principle of Hope* explored at length the role played by the arts, including music, in anticipating Bacon's future *regnum hominis*. Art embodied what Bloch called the 'Not-Yet-Conscious', a prefiguration of the desired future society (1995: 116). Not all art articulated this utopian yearning in the same way, however. Rather, Bloch distinguished between what he called 'abstract' and 'concrete' utopias. Abstract 'utopianising', he wrote, is 'predominantly without solid subject behind it and without relation to the Real-Possible'. As a result, it tends to be 'unworldly' (1995: 145). A concrete utopia, on the other hand, is 'much more centrally turned towards the world: of overtaking the natural course of events' (1995: 12). Whereas an abstract utopia is merely an escapist compensation for present alienation, then, a concrete utopia anticipates future social transformation in a more intellectually and politically mature or plausible way. It does this, wrote Bloch, by combining 'enthusiasm' with 'sobriety': '*enthusiasm* assists sobriety, so that it does not abstractly-immediately foreshorten the *perspective* instead of keeping it on the globe of concrete possibility' (1995: 1368). As Vincent Geoghegan explains, the notion of concrete utopia is thus 'Bloch's reformulation and further development of Marx's concept of praxis, the unity of theory and practice; it is both goal and the actual creation of that goal' (1996: 38). For Bloch, the supreme example of concrete utopianism in music was Beethoven's opera *Fidelio* (1806).

The moment when Leonora releases Florestan from his chains represents a victory over tyranny that stands for '(e)very future storming of the Bastille [...] more than anywhere else music here becomes morning red, militant-religious, whose day becomes as audible as if it were already more than mere hope' (1995: 1103).

Of all the arts, music, wrote Bloch, was 'the supreme art of utopian venturing beyond, whether it drifts or builds' (1995: 1057). He argued that the Western classical tradition best embodied the utopian potential of music, because its tonal harmonies enacted the struggle of the Not-Yet-Conscious to reach fulfilment. Bloch called the sense of forward-movement in Western classical music its 'Faust-energy', which enacted the hope of a future reconciliation of human beings with nature (1995: 1062). By evoking Faust, he was drawing on a key myth in modern Western European societies, which symbolises a conception of history as the progressive struggle of human beings for domination over nature, typified historically, as we have seen, by Francis Bacon. Historical time, in this scenario, is teleological; that is, it moves purposively towards a definite end. Bloch argued that the tonal harmonies of Western European classical music were an enactment of this conception of time as progressive, Faustian striving. His retelling of the story of the natural origins of music from Ovid's *Metamorphoses* (1955: 47-8) illustrates this notion. When water nymphs saved the tree nymph Syrinx from Pan's sexual advances by transforming her into a bunch of reeds in his hands, Pan was so enchanted with the sound of the wind blowing through the reeds that he fashioned them into the first musical instrument, a flute which he named after Syrinx. The first music was therefore an expression of utopian yearning for a lost object, or, in Bloch's words, a 'sounding wishful dream' and a 'pathos of lament for the absent' (1995: 1059).

Significantly, for Bloch, the utopian value of Pan's music lay in its melody, rather than its rhythm or timbre. 'There is an important contrast', he wrote, 'between the syrinx and ritualistic or percussive instruments with their dull, bellowing, howling and rattling noises. Into this ritualistic sound-world was now thrust an instrument giving out a well-organised series of notes. And by uniting syrinx and nymph, Ovid designated the goal towards which the note-sequence – always a tracing of lines in the invisible – is moving. It is something contradictory and utopian, for this flute-playing constitutes the presence of a vanished entity; that which has exceeded the *limit* is regained by this lament and contained in this consolation' (1995: 1059). For Bloch, then, the utopian promise of music lay in the sense of forward-direction created by a melody that resolves itself within the tonal harmonic structure of the Western European classical tradition.

Along with Theodor Adorno, whose work we will come to presently, Bloch was one of the first twentieth-century philosophers to develop a sophisticated theory of musical aesthetics and environmental ethics. Yet the political aspects of Bloch's theory of music are problematic, particularly when measured against his overt political commitments to the Soviet Union. In 1937, he publically defended the Moscow show trials, and only condemned Stalin much later (Geoghegan 1996: 45). The revolutionary or socialist content of the concrete utopias he advocated must be seen in this light. Moreover, the environmentalist implications of Bloch's views on the natural world are also deeply questionable when judged from a more recent environmentalist perspective. As several radical environmentalists have observed, humanist Marxism largely failed to take ecological considerations into account (Grundmann 1991; Eckersley 1992). In particular, the fundamental Marxist assumption that modern technology is not environmentally destructive in its own right, but only because of the way it is organised under capitalism, is seriously flawed. As we have seen, Bloch believed that in a classless, socialist state, modern technology would be efficient and benign, and would liberate all human beings by perfecting their control of the natural world. 'In a framework of different productive relations', wrote Adorno in a similar vein, 'the same technology might be able not to violate, but to help nature realise some of its aims right here on this old earth' (1984: 100).

Yet radical environmentalists argue, to the contrary, that some technologies, nuclear power being the most obvious, pose ecological risks in their own right, irrespective of the way they are organised socially. Nuclear or coal-burning power stations will pose potential health and safety risks even in a classless, socialist society. Moreover, the principle of economic growth, which humanist Marxists assume to be a social good, comes at the cost of continual environmental degradation. Despite these ethical and political objections, however, Bloch's philosophical explorations of aesthetics, nature and technology remain a valuable resource for ecocriticism, particularly because of the way his approach to ideological analysis preserves the social value of the arts. Bloch's critical method combined ideological analysis of the social, political and historical determinants of a work of art with a wider recognition of what he called its 'surplus' meaning, which transcends the particular time and place of its production. By taking this view, he moved beyond the narrow determinism of earlier Marxist theories of culture, in particular Georg Lukács' argument in *History and Class Consciousness* (1922) that the meaning of a work of art can be totally reduced to the underlying ideology of the society that produced it (Lukács 1971: 228-9). Bloch wrote of music, that 'no art has so much surplus over the respective time and ideology in which it exists'. The music of Brahms and Hindemith, for example, not only reflected the 'emotional and goal-world' of the ruling class to which the composers belonged, but also 'the numerous sufferings, the wishes and the spots of light of the oppressed class' (1063). Bloch's notion of 'surplus' thus avoids the reductiveness to which orthodox Marxist critiques of art and culture are prone, and gives a positive and creative role to music, and the arts in general, in cultural life; as such, his theories, despite the reservations outlined here, are a valuable resource for this book.

## 1.2 Postmodern musicology: Susan McClary

Critics of the Faustian conception of linear, progressive time, which we have seen embraced by Ernst Bloch, argue that it has underpinned the destructive effects of capitalism and industrialisation, including the near extinction of many precapitalist societies which are closer to the more organic, cyclical temporal processes of the natural world. In the United States, the radical cultural historian Lewis Mumford argued that the commodification and acceleration of time in industrial society has separated human beings from the organic rhythms of nature (1970: 391). Mumford's late writings fed into the emerging radical environmentalist movement in the 1970s, which came to view Faust no longer as a hero of progress and the fulfilment of Francis Bacon's project of the kingdom of man on Earth, but instead as a bringer of environmental despoliation. Marshall Berman observes that, in this period, the figure of Faust became for many in the counterculture 'the demon who had wrenched mankind out of its primal unity with nature and propelled us along the road to catastrophe' (1982: 82).

These critical attitudes also found their way into the aesthetics of music, which are particularly open to theories about the social and political construction of time. For some art music composers in the

1960s, the teleological structures of Western music, typified by a melody striving towards harmonic resolution, came to be seen as both conceptually flawed and inherently reactionary. For musicologist Leonard Meyer, the new avant-garde of composers, led in the United States by John Cage, began to challenge the Western classical tradition, and, from a socio-political perspective, critiqued thereby the notion of progress which such music appeared to celebrate (Meyer 1967; Ingram 2007a).

By the 1990s, Meyer had influenced the postmodern turn in musicology, which led to further explorations of the relationship between music, including popular music, and changing social constructions of space, time and environment. In *Feminine Endings* (1991), Susan McClary criticised the goal-oriented tonality of Western classical music mainly because she saw it as symbolically reflecting patriarchal attitudes to sexuality. McClary's emphasis on gender politics was, however, part of a wider, if more implicit, social, political and economic critique, according to which the sense of 'tonal striving, climax, and closure' in classical music both reflects and perpetuates the destructive course of Western modernity as a whole, including its damage to the environment. McClary lists amongst what she calls the 'excesses' of modernity, apparently reflected in tonal music, 'the capitalist undermining of more mercantile economic processes, imperialist invasions of 'primitive' societies, scientific quests that replace ecologically grounded philosophies of nature with threats of nuclear destruction, the programs of urban renewal that destroy traditional communities' (McClary 1991: 124). With the onset of modernity, she writes, the 'ideal of culture changes from stability and balance to extravagant, individualistic assertion'. She singles out Beethoven (Ernst Bloch's hero) as exemplar of such destructive tendencies, and, in contrast, advocates alternative forms of music that work in different, ethically healthier ways. Medieval Western music, for example, was characterised by 'relatively noncoercive modal techniques that delight in the present moment, rhythms that are grounded in the physicality and repetitiveness of dance, and the kind of carefully regulated contrapuntal interplay that Renaissance theorists associated with the harmony of the spheres, of nature and humankind, of soul and body'. She also advocates contemporary art music by minimalist composers such as Philip Glass and Steve Reich, and feminist composers such as Laurie Anderson and Janika Vandervelde, because it either rejects teleological structures altogether, or plays them off against cyclical ones. The latter McClary sees as more environmentally benign than linear structures, because, as she puts it, they 'resonate with the patterns of nature: seasonal yet timeless, always fascinatingly different and yet always the same' (119).

Before discussing the relevance of these speculations to an eco-aesthetics and ethics of popular music, I will examine a second critique of Faustian, teleological form in music which emerged from French post-structuralist philosophy.

## 1.3 Post-structuralism: Deleuze and Guattari

Like Susan McClary, the French post-structuralist philosophers Gilles Deleuze and Félix Guattari also extended the concept of anti-teleology in music in ecophilosophical directions. The title of their book *A Thousand Plateaus* derives from Gregory Bateson's work on Balinese culture, in which the British anthropologist developed the concept of the 'plateau' as what Deleuze and Guattari call 'a continuous, self-vibrating region of intensities whose development avoids any orientation towards a culmination point or external end' (1987: 22). The rhizome is their key image for this notion of decentralised, non-hierarchical and process-sual form. Rhizomes are plants, such as a potato and ginger, which differ from trees (or what Deleuze and Guattari call 'arborescent' structures) in that they have no central root structure, and grow by lateral and non-linear movements. Deleuze and Guattari apply their concept of 'rhizomatics' to both aesthetics and political theory, by mapping it onto other sets of oppositions: 'deterritorialisation' versus 'reterritorialisation', the 'nomadic' versus the 'sedentary', and 'smooth' versus 'striated' space. Put briefly, the former processes, associated with the rhizome, are ongoing 'lines of flight' away from established, fixed or regular order, while the latter attempt to assert stasis, order and fixity.

Deleuze and Guattari's analysis of music starts from the notion of the 'refrain'; that is, a small rhythmic or melodic pattern, examples of which are lullabies, folk songs and birdsong. They argue that music is a constant interplay between 'deterritorialising' and 'reterritorialising' the refrain. Highly organised or rule-bound musical forms such as the sonata, or strict metres measured by bar lines, are examples of territo-rialisation. These forms, they write, are 'compartmentalised, centralised, and hierarchised in relation to one another' (338). In contrast, deterritorialised music takes on more unpredictable and non-linear forms, or brings out the sonorous qualities of the musical material in itself.

In *What Is Philosophy?*, Deleuze and Guattari explore an explicit conceptual link between music and the natural world, by claiming that music is homologous to the biologist Jacob von Uexküll's 'melodic, polyphonic, and contrapuntal conception of Nature'. The natural world is made up of complex, non-teleological, mutually enhancing interactions. In Deleuze and Guattari's example: 'The tick is organically constructed in such a way that it finds its counterpoint in any mammal whatever that passes below its branch, as oak leaves arranged in the form of tiles find their counterpoint in the raindrops that stream over them' (1994: 185). If 'nature is like art', they continue, 'this is always because it combines these two living elements in every way: House and Universe, *Heimlich* and *Unheimlich*, territory and deterritorialisation, finite melodic compounds and the great infinite plane of composition, the small and large refrain' (1994: 186).

Deleuze and Guattari's conception of music shows marked aesthetic preferences, which become clear as they map the distinction between deterritorialisation and reterritorialisation onto that made by the French composer Pierre Boulez between 'smooth' and 'striated' space. Smooth space, they write, 'is occupied by intensities, wind and noise, forces, and sonorous and tactile qualities, as in the desert, steppe, or ice. The creaking of ice and the song of sands'. Striated space, in contrast, is the space of human order and control, 'canopied by the sky as measure and by the measurable visual qualities deriving from it' (Deleuze and Guattari 1987: 479). In a smooth space, 'one occupies without counting', whereas in a striated space-time, 'one counts in order to occupy' (1987: 477).

From an ecophilosophical perspective, this opposition to 'counting' reveals the anti-rationalist element in Deleuze and Guattari's philosophy, which has important implications for their attitude to environmental ethics. With Michel Foucault, Deleuze led the turn towards Nietzsche in post-structuralist philosophy in the 1960s. According to this line of argument, truth-claims are merely assertions of power. Nietzsche's method, for Deleuze, was 'as follows: relating a concept to the will to power in order to make it the symptom of a will without which it could not even be thought (nor the feeling experienced, nor the action undertaken)' (1983: 78). In *What is Philosophy?*, Deleuze and

Guattari credit Nietzsche for 'making us understand, thought is creation, not will to truth' (1994: 54). A philosophical idea is to be valued, then, for other criteria than its truth. Philosophy, they write, 'does not consist in knowing and is not inspired by truth. Rather, it is categories like Interesting, Remarkable, or Important that determine success or failure' (1994: 82).

Yet this turn towards extreme relativism and constructivism in post-structuralist philosophy is highly debatable, to say the least. As Jürgen Habermas observes, the Nietzschean position is both self-refuting and self-defeating as a critical method. He asks: 'once all predicates concerning validity are devalued, once it is power and not validity claims that is expressed in value appraisals – by what criterion shall critique still be able to propose discriminations? It must at least be able to discriminate between a power that *deserves* to be esteemed and one that *deserves* to be devalued' (1990: 125).

In the case of Deleuze and Guattari, power is esteemed if it is anarchic, and devalued if it is capitalist or derived from the State. Crucially for environmental ethics, such political values inform Deleuze and Guattari's epistemological position on science and rationality. Following Michel Serres, they distinguish between a 'royal' or 'legal' science, which reinforces the interests of the State, and a 'nomad' or 'minor' science, which is subversive of those interests. This pairing also maps onto Boulez's distinction between '*striated* (metric)' and '*smooth* (vectorial, projective, or topological)' space, mentioned earlier (Deleuze and Guattari 1987: 361-2). State science, they write, 'continually imposes its form of sovereignty on the inventions of nomad science. State science retains of nomad science only what it can appropriate; it turns the rest into a set of strictly limited formulas without any real scientific status, or else simply represses and bans it' (362). Archimedes' treatment by the Roman state stands as a central instance of the repressiveness of State science: he is an example of 'savants whom State science used only after restraining or disciplining them, after repressing their social or political conceptions' (363).

This post-structuralist critique of Enlightenment notions of science and reason limits the usefulness of Deleuze and Guattari for ecocriticism. As Ursula K. Heise writes, 'if the context out of which scientific research emerges is shaped by certain values, it does not necessarily follow that the results of this research will lend support to these values, a distinction that few cultural analyses of science bother to make' (1997: 4ff). Science, then, does not reflect political or ideological interests in the simplistic way described by Deleuze and Guattari. Gross and Levitt take the argument against post-structuralist critiques of science further. 'The relativism of cultural constructivist doctrine', they write, 'is the perfect tool for discounting science as biased or corrupt if and when it inconveniences one's political program' (Gross and Levitt 1994: 162). In this way, the inclination 'to regard heterodoxy, per se, as intrinsically valuable persists in the oppositional subculture of our own day, in its attitude and in its choice of philosophical godfathers' (222-3).

## 1.4 Critical realism and political ecology

Instead of the post-structuralism embraced by Deleuze and Guattari, this book favours the competing paradigm of critical realism as a sounder philosophical basis for ecocriticism. In particular, the 'scientific realism' element within critical realism provides a more defensible position on the epistemological disputes over reason, science and the social constructedness of nature which are central to ecocritical debates.

Critical realists argue that human conceptions of 'nature' are multiple and varied, and in this sense, that nature is a socially produced, discursive construct. However, they also argue that nature exists external to human beings, and is objectively knowable. As political scientist Tim Forsyth puts it, critical realism combines 'epistemological scepticism, and a commitment to an ontological realism of underlying biophysical processes. In other words, the belief that biophysical reality is "externally real" to human experience, notwithstanding the fact that all knowledge of such reality is partial and socially constructed'. Critical realism thus forms the basis for a realist political ecology, which, he continues, seeks 'to integrate political awareness of environmental conflicts with a realist understanding of environmental change' (2001: 147). In doing so, critical realism rejects the extreme scepticism towards science and truth towards which post-structuralist philosophers such as Deleuze and Guattari tend to gravitate.

The causal realist, Christopher Norris writes, believes 'as a matter of inference to the best explanation – that science has achieved genuine advances in our knowledge of a (largely) mind-independent and language-independent reality' (1997: vii). By advocating scientific method as the most effective means of investigating the truth-value of statements, critical realism thus rejects the extreme relativism of post-structuralist theorists such as Deleuze and Guattari, who assume that all human knowledge is 'culturally embedded', and therefore epistemologically relative. The confusion of postmodernist relativism, Kate Soper points out, lies 'in supposing that because we can only refer in discourse to an extra-discursive order of reality, discourse itself constructs that reality' (1996: 30). Critical realism corrects this misconception by positing a natural world 'whose structures and processes are independent of human activity (in the sense that they are not a humanly created product) and whose forces and causal powers are the condition of and constraint upon any human practice or technological activity, however Promethean in ambition' (31).

## 1.5 Popular music and structural homology

Each theorist encountered in this chapter has advocated particular styles of music as best expressing the eco-utopian imagination. For Ernst Bloch, Beethoven best embodied the Marxist, Faustian hopes in which he placed his faith; overall, the Western 'classical' music tradition, including twentieth-century modernism, was the sole repository of such cultural value. For Susan McClary, conversely, pre- and post-classical styles, such as pre-Renaissance modal music and minimalism, challenged Faustian hubris. What unites these disparate ethical-aesthetic positions is a notion of 'structural homology', whereby the formal structures of music are seen as homologous to the social and political structures of human society. For Bloch, tonal harmonic progression is homologous to Faustian progress; for McClary, cyclical or modal forms are homologous to a society in which Faustian notions of progress have broken down; for Deleuze and Guattari, rhizomatic, deterritorialised musical forms are homologous to revolutionary anarchist modes of social and political practice. In concluding this analysis of these rival theories, their common assumption of a structural homology between musical and social structures needs to be addressed.

As musicologist Allan F. Moore observes, musicology has not yet satisfactorily determined 'the mechanism(s) by which social relationships and cultural practices are embodied in musical sounds' (2001: 6). Indeed, the relationship between the two levels is often merely one of loose analogy. Adorno argued that tonal music was homologous to the capitalist society that produced it, yet what it literally has to do with the economics of labour is uncertain. The same can be said for the correspondences between musical form and socio-political structures explored in this chapter.

Enrico Fubini clarifies the role of structural homology in theories of music and society by questioning the assertion of a correspondence between music and social structure in Marxist aesthetics. 'How is any correspondence between structure and superstructure supposed to be manifested in the language of music?' he asks. 'What degree of independence does music enjoy in relation to the economic structure of society?' (Fubini 1987: 436). He concludes that the language of music is ideologically neutral, so that any assertion of a correspondence between musical style and social structure is a social and historical construction, and therefore necessarily loose and provisional. We may conclude, then, that the formal correspondences that the theorists discussed in this chapter find between music and particular social structures, or between music and particular concepts in environmental ethics, are ultimately matters of ethical and aesthetic preference, rather than of necessity. It is not that Bloch's advocacy of linear teleology in music is less 'ecological' than McClary's antipathy towards it. Nor is a particular guitar riff, drum pattern or vocal harmony more or less ecological than any other. All such claims are conceptual constructions. The method employed in this book, therefore, is to interpret the popular music made in the United States since 1960 in the light of the theoretical models explored in this chapter, while acknowledging the partiality and fallibility of each approach. This is preferable to turning the whole study into a 'Blochian' or 'Deleuzian' analysis of music in a way that merely applies a single theoretical template to the material at hand, however flawed or questionable that approach is philosophically.

This chapter has explored three broad theories of music and society in relation to their position on environmental ethics. Their specific relevance to a consideration of popular music, as opposed to music in general, will be the subject of the next chapter.

# Chapter Two

# Popular music and eco-aesthetics

## 2.1 Political modernism and popular music

Theodor Adorno's belief in the revolutionary importance of modernist music came from his Marxist sense that the formal structures of the Western tonal tradition were an inadequate response to the catastrophe of twentieth-century history. This chapter explores the implications of Adorno's theories for an eco-aesthetics of popular music, and examines recent postmodernist challenges to his position.

Arguing from the notion of structural homology that we examined at the end of the previous chapter, Adorno believed that the tonal system on which both Western classical and popular music is based reflected the capitalist values of the society that invented it. As he put it, tonality 'owes the dignity of its closed and exclusive system to mercantile society, whose own dynamics stress totality and demand that the elements of tonality correspond to these dynamics on the most basic functional level' (Adorno 1948: 11). By the twentieth century, however, tonality had '*disintegrated*', so that every tonal chord 'has a meaning which we can no longer comprehend' (italics in the original) (Buck-Morss 1977: 224n.57). For Adorno, only the dissonances and discontinuities of modernist music could truly reveal the contradictions of capitalist society, and therefore articulate music's latent utopian potential. Adorno was influenced in such thinking by Bertolt Brecht's concept of *Verfremdungseffekt*, which is usually translated as either the 'defamiliarisation' or the 'estrangement' effect. 'Perspectives must be fashioned', Adorno wrote in *Minima Moralia*, 'that displace and estrange the world, reveal it to be with its rifts and crevices, as indigent and distorted as it will appear one day in the messianic light' (1974: 247). Music's radical utopianism was best embodied by serialism, the technique invented by the Austrian composer Arnold Schoenberg of composing music using all twelve notes of the diatonic scale in succession. The atonality and dissonance of Schoenberg's music made the act of listening work, rather than mere passive consumption (Adorno 2002: 632). In doing so, serialism thereby fulfilled the Enlightenment project of reason in the face of what Adorno saw as the irrationalism of mass culture.

Adorno's antipathy to popular music is well known. He opposed music as entertainment, because it 'no longer does anything but confirm, repeat and reinforce the psychological debasement ultimately wrought in people by the way society is set up' (Adorno 1948: 225). In *The Dialectic of Enlightenment*, he argued with Max Horkheimer that the 'culture industries' merely reinforce the capitalist system and prevent the emergence of a liberatory consciousness. Pleasure, Adorno and Horkheimer wrote, 'is flight; not, as is asserted, flight from a wretched reality, but from the last remaining thought of resistance' (Horkheimer and Adorno 1973: 144). Modernist music, on the other hand, provided a critical perspective on society, and could therefore play a vital role in the socialist project of liberation from capitalism.

The issue of musical form was central to Adorno's antipathy towards popular music. He argued that the success of popular music as social control derived from the formal characteristics it shared with capitalist processes of industrial standardisation; namely, part-interchangeability and 'pseudo-individualisation', in which deviations from the stylistic norms imposed by the culture industries are themselves standardised in order to give an illusion or 'halo of free choice' (Adorno 2002: 445). Jazz was falsely liberating, then, in that, while its improvised forms seemed to promise an ideal of individual freedom, this was merely deceptive, because it ultimately relied on stereotypical and commercial musical forms (478).

Adorno's blanket dismissal of popular music is a major weakness in his theorising about music. His criticism of jazz, ignorance of non-European musics and exclusive emphasis on the value of composers, scores and predetermined theories rather than on oral performance traditions, all indicate the Eurocentric bias and class snobbery that marred his aesthetic judgments. As he himself pointed out in his essay 'On Jazz', although popular songs may be standardised at the level of melody and harmony, they are less so in terms of timbre, rhythm or sonority (Adorno 2002: 470). Indeed, popular music has been constantly innovative in these areas. In denigrating these aspects of music, then, Adorno was basing his aesthetic judgments on the culture-specific values of Western classical music. Moreover, his Puritanical attitude to pleasure lacked an understanding of the complex ways in which, in the reception of popular music, pleasure and political resistance can be mutually reinforcing, as in the musics of the African diaspora, for example. Indeed, Adorno had no faith in the positive aspects of collective action; instead, his experiences as a refugee from Nazism had convinced him that all appeals to collectivity meant tyranny.

The aesthetics of political modernism, which informed Adorno's negative attitude to popular music, remain influential in music criticism. Jacques Attali's *Noise* reiterates his negative view of the music industry as a form of top-down manipulation and pacification. In what he calls the era of Repetition, writes Attali, the 'network is no longer a form of sociality, an opportunity for spectators to meet and communicate, but rather a tool making the individualised stockpiling of music possible on a huge scale. Here again, the new network first appears in music as the herald of a new stage in the organisation of capitalism, that of the repetitive mass production of all social relations' (1985: 32). Writing from a more ecologically concerned perspective, composer and phonographer David Dunn also argues that popular music is fatally complicit in global capitalism, so that only avant-garde experimentalism can resist its environmentally destructive forces (1988: 9). Despite his call to 'refuse value judgments', John Cage's aesthetics continue to be influential in such endorsements of the resistive power of the avant-garde (Cage 1968: 14).

Ironically, given the music's commercial basis, many commentators on rock music, both musicians, critics and fans alike, have also reflected Adorno's modernist values, as they seek to differentiate rock from the supposedly more commercially degraded forms of popular music. Speaking in 1971, Captain Beefheart explained his concept of rhythm: 'I'm tired of lullabyes [sic], like the Beatles. I heard "Lullaby of Broadway" when I was a baby, and I still hear it now, and I'm still a baby' (Chorush 1971: 2). He wanted, he said, 'music without a lullaby', in order to 'de-materialise the catatonia' (Bamberger 1999: 72). Listening to music, Beefheart argued, should be a participatory action, not an easy act of commodity consumption. Frank Zappa voiced similar opinions; indeed, Ben Watson's *The Negative Dialectics of Poodle Play* explicitly reads Zappa as an exemplar of Adorno's aesthetic values (Watson 1994).

The values of political modernism are also at the heart of the writings of Deleuze and Guattari, which we examined in the previous chapter. Indeed, popular music plays an ambiguous role in their aesthetic preferences. Although they are most interested in art composers such as John Cage, Pierre Boulez and Olivier Messiaen, who are presented as the main exemplars of rhizomatic, non-teleological music, there are moments when Deleuze and Guattari appear more open to the aesthetic possibilities of popular music (1987: 267). Nevertheless, the pair did not write about popular music in depth, so other music scholars have extended and applied their ideas more fully in that direction. Such critiques tend to favour popular music which is defined as 'deterritorialised', as opposed to the supposed territorialisation of mainstream pop. Greg Hainge, for example, draws a distinction between 'Pop' and 'popular'. For him, the music of the British trip hop artist Tricky 'performs rhythmic and linguistic deterritorialisation at odds with the ordered, predictable and punctual structures of pop'. The vocal and guitar lines 'have a deterritorialising effect as they seem to warp and bend the regular, ordered meter of the pulsed beat' (2004: 44). In this context, then, deterritorialisation is effectively another word for 'groove'. Unfortunately, however, Hainge does not give an example of the popular music of which he disapproves. Instead, he falls back on the argument that popular music is automatically bad if it is commercially motivated. 'If the motivation behind the territorial expression of Pop issues not from an internal imperative but from an external imperative, an axiomatic of the strata of Capital', he writes, 'it cannot become auto-objective and will instead become an auxiliary of the apparatuses of capture of the Capitalist machine' (51). Deleuzian music theory here recalls Adorno's negative view of popular music as inevitably valueless.

In contrast to the aesthetic values of political modernism outlined so far in this chapter, the postmodernist musicology of Susan McClary and others allows for a more aesthetically pluralist approach to music, and raises the question of how the theories of music as eco-utopia addressed in the previous chapter can be applied specifically to popular, as opposed to 'art', music.

## 2.2 Postmodern aesthetics and popular music

Although he was as hostile to American popular music as Adorno, Ernst Bloch's interest in the pleasures of melody and tonality ultimately opened the way for a more inclusive aesthetic, and so provide a useful bridge to an exploration of more recent critiques of musical modernism. Bloch's attitudes to popular music will now be examined. His lack of enthusiasm for popular music came from his preference for melody over percussive rhythm. Although he did not write at length on the subject, his remarks on the popular American dance crazes of the 1930s, such as the jitterbug and boogie-woogie, made his aesthetic position clear. These dances were, he wrote, 'imbecility gone wild, with a corresponding howling which provides the so to speak musical accompaniment [...] Man is to be soiled and his brain emptied; he has even less idea amongst his exploiters where he stands, for whom he is grafting, what he is being sent off to die for' (Bloch 1995: 394). Bloch's antipathy to popular American dance culture was thus racially marked: he considered the jitterbug and boogie-woogie, like the African tribal dances on which they were based, as cultural atavisms that ran counter to the necessarily progressive direction of history, which involved the taming of nature through reason and modern technology. Music was for him the art which rises above 'nature's spell', rather than surrenders to it (1083). Accordingly, for him, the values of white, Western European high culture best embodied his utopian hopes.

Despite Bloch's negative views on popular music, however, he was not hostile to popular culture *per se*. The title of the section in *The Principle of Hope* that discussed the utopian potential of fairs and circuses is revealing in this respect: 'Better Castles in the Air' (352ff). Crucially, then, even the degraded forms of abstract utopia were for Bloch preferable to a cultural pessimism that merely confirms the status quo. Moreover, despite his hatred of American popular music, Bloch's theory of tonality as utopian yearning can be turned into a resource for a more positive evaluation of popular music, and thus open the way for the critical appreciation of a wider range of music than that allowed for by both his own and Adorno's aesthetic preferences.

Tonal music creates a sense of forward motion away from, and then back towards, the 'home' of the final keynote. As musicologist Victor Zuckerkandl observed, 'Arrival, departure, further removal, return home; it appears that intervals, if they are heard in accordance with their dynamic meaning – if they are heard musically, that is – cannot be heard otherwise than as phases of a motion, as steps' (1969: 93). Like Bloch, Zuckerkandl himself was concerned with the aesthetic value of Western high cultural music. Nevertheless, his theory of tonality can be extended to popular musical forms as well. Even though popular music is simpler, melodically and harmonically, than a Beethoven sonata, it still enacts a similar sense of forward-striving, and of playing with a listener's expectations for closure. The difference between popular and 'art' styles, in this sense, is one of degree rather than kind. To put it briefly: there is a place within eco-aesthetics for the pleasures of a good tune.

Moreover, the idea that there are subtleties in both the form and the performance of most popular songs not only makes Adorno's antipathy to popular music untenable, but also makes Deleuze and Guattari's notion of the refrain critically superfluous. 'In real life', writes songwriter Jimmy Webb, 'music breathes and follows its own path. The metronymic cadences of classical rhythm must give way, certain notes must be elongated, space must be allocated for the singer to take air, long musical phrases may call for more metrical feet in one line than another, other musical lines will be abrupt and attenuated, etc. In other words there is a *nonregimented* quality to most non-military melody' (italics in the original) (1988: 89). Put into Deleuzian language, most popular songs are 'deterritorialised', although some tunes are more or less unpredictable than others. In light of this idea, the aesthetically pluralist approach taken in this book is intended to be open to the multiplicity of musical forms that musicians and singers create.

The utopian value of popular music can also be defended from a sociological perspective in a way that provides another important theoretical pathway for this book. In the 1970s, the British school of Cultural Studies placed a new emphasis on the reception of popular music, and therefore challenged the formalist aesthetics of Adorno by providing insights into the way audiences create meanings for themselves around different types of music. In his essay 'Entertainment as Utopia', Richard Dyer analysed entertainment in general as offering 'the image of "something better" to escape into, or something we want deeply that our day-to-day lives don't provide. Alternatives, hopes, wishes – these are the stuff of utopia, the sense that things could be better, that something other than what is can be imagined and maybe realised' (1985: 222). Dyer's main example of the utopian potential of popular entertainment was the Hollywood musical, which, he argued, provides imaginary solutions to perceived 'social tensions, inadequacies, or absences': energy for exhaustion, abundance for scarcity, intensity for dreariness, transparency for manipulation, and community for fragmentation. In this way, the musical provides 'temporary answers to the inadequacies of the society which is being escaped from through entertainment' (222).

Dyer argues that this imaginative process has its political limitations, however, in that entertainment tends to ignore those needs which may only be satisfied by a radically different social order. The answers provided by entertainment thus tend only to apply to needs that capitalism itself promises to meet: the need for abundance is answered by consumerism, energy and intensity by personal freedom and individualism, and transparency by freedom of speech. Popular culture, then, tends to limit and contain the very utopian desires it expresses. In this sense, Dyer recalls Bloch's notion of 'abstract utopia' discussed in the previous chapter. Like Bloch, however, he also asserts that the utopianism of popular entertainment can be ideologically subversive of the status quo. 'To be effective', Dyer writes, 'the utopian sensibility has to take off from the real experiences of the audience. Yet to do this, to draw attention to the gap between what is and what could be, is, ideologically speaking, playing with fire' (228-9). In contrast to Adorno, then, Dyer considers the audience for popular culture as active rather than passive; audience reception can embody progressive desires for social change.

For sociologist Jason Toynbee, the utopian potential of popular music resides in the context of its production rather than its reception. Musicians, he writes, demonstrate 'in a limited, but none the less substantive, fashion the transforming power of human agency, first as producers of desire for a better life, second as exemplars of autonomous action' (2000: xiii). The work of popular musicians particularly embodies values such as participative democracy and individual self-development. Toynbee's point is illustrated by Tuli Kupferberg of The Fugs, who articulated the prefigurative promise of popular music in 1968 in an essay entitled 'The Coming Catastrophic Age of Leisure': 'Play is as good as work. Work has been defined as something you dislike doing. Fuck that. Do the Beatles work? Who cares. We like what they do' (Berke 1969: 85). However, Toynbee adds the important rider that we should not expect musicians 'to change the world on their own' (xiii).

Having established the utopian potential of popular music in both its production and reception, the next step in developing an eco-aesthetics of popular music is to move beyond the narrow anthropocentric concerns of Toynbee and Dyer. Can popular music be a utopian prefiguration of a relatively benign relationship between human beings and the natural world, which includes ethical consideration for non-human entities? Those musicians and ecophilosophers who make such claims tend to focus on two related phenomena: the rhythmic 'groove' of dance music, and the public spaces and events in which such music is performed. Anthropologist Steven Feld describes 'groove' as the 'varieties of out-of-timeness (process) and out-of-tuneness (texture)' which combine to 'generate music's vital force, as well as guaranteeing the active qualities of participation' (Feld 1994: 120). Musicologist Charles Keil similarly defines 'groove' as 'participatory discrepancies', and has developed an explicitly ecological theory based on it (Keil and Campbell 2006). For Keil, as we will see in more detail in Chapter 10 on world music, music with a groove not only liberates the body and mind of its participants, but can also establish a sense of community that is imaginatively extended to include not only musicians and their audience, but the natural world itself. In later chapters, we will see how musicians as diverse as David Rothenberg and ex-Grateful Dead drummer Mickey Hart similarly affirm the potential of rhythmical groove to foster ecological consciousness.

In *Constructing Musical Healing*, the British composer and music therapist June Boyce-Tillman attempts a synthesis of Jungianism, Hinduism and the Human Potential Movement, the school of ego psychology that promotes creativity and play as steps towards 'self-actualisation'. Like Suzi Gablik, she argues that the 'distrust of rationalism, awareness of the feminine and ecological, quest for a post-materialist quality of life, return to spirituality, sense of dispersal and disillusionment with Utopia that characterises post-modern sensibility are strands leading to a re-evaluation of the role of music' (Boyce-Tillman 2000: 11). She endorses postmodernist aesthetics because of their renewed emphasis on what she calls 'human-centred' and 'contextualised' approaches to music (in other words, live performance and audience participation), alongside more formal approaches based on notation and scoring.

Aesthetically, the icon of Boyce-Tillman's Green Christian ecofeminism is the twelfth-century German abbess and composer Hildegaard of Bingen. However, she also advocates New Age music, because its tonal familiarities can satisfy a listener's therapeutic need for 'nurturance'. New Age music, she writes, is 'maternal, designed to relax, to affirm people where they are – to nurture them in the same way as a mother might feed a child' (54-5). In this way, Boyce-Tillman distances herself from the exclusive emphasis on 'challenge' that has dominated the aesthetics of Western classical music from the Enlightenment, through Romanticism, to the twentieth-century modernism of Adorno. As a result, she refuses to dismiss entertainment music, including amateur and folk musics, as merely shallow or reactionary.

Boyce-Tillman's musical aesthetics can also be seen as an extension of Pythagorean notions of music. Enrico Fubini places ancient Greek notions of 'catharsis', or the therapeutic power of music to purify the soul, into two categories: 'allopathic' and 'homeopathic'. The former is identified with the Pythagorean philosopher Damon, who exerted an important influence on Plato; the latter is a position associated with Aristotle. For Damon, Fubini explains

> music has the power not only to educate the hearts and minds of men but also to correct their proclivities for wrongful behaviour. Correction can be brought about through the influence of a kind of music that imitates the particular virtue which it is desired to inculcate, and which at the same time annuls the earlier proclivity to vice. In this case the catharsis is allopathic; it is induced by an imitation of the virtue which is opposed to the corresponding vice. On the other hand, Aristotelian catharsis should probably be termed homeopathic, since any correction to the vices can be ensured by a portrayal of the very vice from which it is desired we should be freed. By this means the power of these vices within us is reduced, and, by listening to the kind of music that portrays the feelings that beset us – such as 'pity or fear or any other emotion' – our minds are 'cleansed and purified' from their influence and we are restored as if we had 'undergone curative and purifying treatment' (Fubini 1987: 29)

In this light, the music advocated by Boyce-Tillman as nurturance for the mind and body works on the listener as allopathic catharsis; the same may be said for Bloch's notion of the utopian power of hope in music. Good music, in this Platonist sense, reflects the ethical good to which human beings should aspire.

The emphasis on nurturance rather than challenge in the post-modernist aesthetics advocated by June Boyce-Tillman tends to reverse the modernist values espoused by Adorno and his followers. Musicologists Shepherd and Wicke state clearly the modernist position that the former rejects. 'People may, on the one hand, be easily positioned by familiar music that they desire', they write. 'In this kind of situation, little more than the reaffirmation of previous patterns of awareness may be at stake. However, in other situations, the individual may make a conscious choice to engage in a dialogue with new and unfamiliar musics that require a readjustment of the self' (Shepherd and Wicke 1997: 177). In practice, however, the oppositions that frame these debates, between 'unfamiliarity' and 'familiarity', and 'art' and 'entertainment', are neither fixed nor constant. With repeated listening, the unfamiliar becomes familiar, and what was initially challenging becomes reassuring. Moreover, the potential feeling of cultural superiority for those willing to take on modernist difficulty may itself be a form of emotional reassurance. The relationship between art and entertainment is thus best thought of as a continuum rather than an absolute dichotomy. Listening to music can involve both a sense of nurturance that comes from the recognition of familiar patterns, and the challenge of the unfamiliar and unexpected. Modernists tend to favour the latter, and unfairly or disingenuously denigrate the pleasures of the former. 'When well-balanced', Boyce-Tillman concludes, 'challenge and nurture together lead to empowerment and the two work in tandem, especially in person-centred models of learning' (55). There is a place for both nurturance and challenge in a healthy musical culture. This post-modernist embrace of popular music is the critical position adopted in the rest of this book.

# Chapter Three

# Popular music and 'nature'

American musicians and composers have evoked many differing conceptions of the natural world when discussing their musical compositions and performances. When considering popular music in relation to ideas about 'nature', two particular conceptions are relevant. Firstly, advocates of folk music claim that authenticity in music derives from the performer or style of music being close to nature. This idea of nature as a source of ethical, and therefore aesthetic value, will be explored in the first part of this chapter. The second idea of 'nature' relates to the question of how the natural world as been represented in popular music. This question will be explored in relation to ecocritical theories of the pastoral mode.

## 3.1 Popular music, authenticity and ethical naturalism

Ethical naturalism is the belief that ethical values may be derived from the principles of the natural world. It is at the centre of what historian Catherine Albanese calls the 'harmonial ethic', according to which conformity to the laws of nature will lead to a good society (1990: 83). The philosopher Jean-Jacques Rousseau exemplified this ethic in European Romantic thought. Albanese traces the harmonial ethic through North American culture from the early Republic and the New England Transcendentalists to the New Age movement of the early 1970s.

Ethical naturalism has also informed much ecocriticism, which has tended to perpetuate the assumption that the natural world is superior to human culture, and consequently that a conscious return to nature is necessary for the betterment of human society. The British ecocritic Jonathan Bate writes approvingly that 'the controlling myth' of ecopoetics is 'a Rousseauesque story about imagining a state of nature prior to the fall into property, into equality and into the city' (2000: 266). Human culture is also a fall from nature for David Rothenberg, who writes that 'while we are created from nature, we are somehow cast out by our wily, civilised ways. And that is why there is a perennial aesthetic pull of the earthy, the natural, the green, the living' (2002: 91).

The Romantic notion that proximity to nature means moral worth has shaped some of the fascination with folk music in American culture. Folk song and dance, writes ethnomusicologist Philip Bohlman, 'create a narrative space for the past to live in the present' (2002: 71). This relationship with the past is also a relationship with the natural world and the rural, and can take two forms: a conservative desire to return to old ways or a more progressive desire to reclaim the best of the past for what may be useful in the future. Debates in folk music circles between 'revivalism' and 'fusion' reflect these alternatives: the former involves the purist search for static 'roots', while the latter involves experimentation, innovation and an embrace of musical hybridity. Yet these positions are not always mutually exclusive; the recent boom in world music, which is the subject of chapter 10, celebrates both new musical fusions and the older, traditional styles from which they emerge.

For some commentators, authentic folk music is synonymous with the closeness to the natural world of the cultures that produce it. Lone Wolf Circles notes 'the decay of substance in most "popular" music, far from its folk roots' (1991: 55). His view of musical authenticity recalls attitudes inherited from the Old Left critique of mass culture, according to which the commodification of music is a wholly negative process that degrades performers, audience and the music itself. For Lone Wolf Circles, folk music transcends commodification, and is therefore the appropriate form of cultural expression for deep ecology. Indeed, he argues that folk music is a cultural survival of the biocentric tribalism that he believes characterised premodern, hunter-gatherer societies, and is therefore an instructive model for contemporary ideas about bioregionalism. His comments, made in 1991, are worth quoting at length:

> The original (archaic) meaning of 'folk' is tribe. Folk music defined the form and biorhythm of the tribe. Tribal identity was maintained, fostering the cohesion necessary for the hunter/gatherers' survival. Each song held a seed, shared only with the trust of the clan. It bore fruit as wisdom, courage, and allegiance to the tribe's sacred view. But even more important than clan bonding, song-seeds renewed the bond to their home reinforcing the biocentrism that allowed them to live vigorously yet harmoniously on their land, inspiring their allegiance to their region (55).

Here, Lone Wolf Circles is in a long line of Americans who have idealised the 'folk' for their own ideological purposes. The semi-mythic past he creates is founded on unsubstantiated anthropological assertions about tribal societies, which he uses to justify his deep ecologist agenda of opposition to urbanism and industrialisation. The biocentric reverence for the sanctity of all creation is, he continues, 'the unified power that strikes technology down with a consuming rust, spinning sand in a display of revolutionary circles' (53). The implication that 'primal' peoples do not use technology shows the Romantic naturalism that informs his New Age position; the proliferation of organic metaphors ('song-seeds' that bear 'fruit') confirms it. On the other hand, if he is condemning, not all technology, but the use of iron in particular (hence the reference to 'consuming rust'), he is merely dismissing over two millennia of human history as an ecological mistake.

Lone Wolf Circles' view of tribal societies fails to take into account recent findings in anthropology which cast doubt on the figure of the 'ecological Indian'. The work of Robert Brightman on the Rock Cree people, for example, concludes that their religious belief in animal spirits led not to a conservationist ethic, but to overhunting, and consequently the excessive destruction of wildlife. Shepard Krech III reaches a similar conclusion about many other American Indian tribes (Lewis 1992: 63; Krech 1999). In the absence of such counter-arguments, Lone Wolf Circles continues the Romantic view of so-called 'primitive' peoples, who, as art historian Marianna Torgovnick observes, are seen as 'by turns gentle, in tune with nature, paradisal, ideal – or violent, in need of control; what we should emulate or, alternately, what we should fear; noble savages or cannibals' (1990: 3). This Romantic cult of the primitive considers primal peoples as exemplars of a closeness to nature that the urbanised West has lost. Geographer Martyn W. Lewis is critical of those radical environmentalists who believe 'that we afford respect to primal peoples by elevating them to the status of nonalienated humanity in its essence, at one with earth and nature. In reality, this procedure denies them their own existence, making them instead a dumping ground for our own fears and longings' (1992: 243-4).

In the United States, the Romantic cult of the folk primitive reflects the inequalities of class and race relations in that country, and has informed discourses on both Appalachian folk music, country and the blues. This naturalist aesthetics of folk music obscures its social and historical constructedness. Indeed, this was the main reason for Theodor Adorno's objection to the conscious revival of folk musics in early twentieth-century Europe. In his essay 'On the Social Situation of Music' (1932), he criticised Igor Stravinsky, because in his music the 'musically immanent correction of alienation is rather sought through regression to older, totally pre-bourgeois musical forms, within which an effort is made to affirm an original nature of music [...]' (Adorno 2002: 403). To claim that folk music is a natural, organic form of music is to deny that all music is a cultural artefact produced under specific historical conditions. Adorno thus approved of Béla Bartók, because his use of traditional folk forms resembled the modernist techniques of defamiliarisation that made the constructedness of the artwork visible. However, he considered the revival of folk musics an evasion of the social and political conditions of the present; the return to traditional folk forms merely affirmed a conservative view of society. Although he rejected the radical potential of folk music in this way, Adorno's point about the danger of viewing folk music as natural, and hence the need to acknowledge the historical specificity and cultural constructedness of musical forms, is an important one.

The notion of 'organic' form is a related aesthetic consequence of ethical naturalism. Writing of literature, ecocritic Dana Phillips notes that 'the resemblance between "organic" literary forms (if there are any) and organisms or ecosystems is entirely negligible, and is therefore devoid of diagnostic significance'. He goes on to observe that the necessity of 'revaluing nature' does not therefore depend on 'the possibility or impossibility of resembling nature' (2003: 144). Organicist or naturalist principles may produce interesting works of art, then, but are not a necessary prerequisite for artistic responses to ecological crisis.

From this perspective, the embrace of naturalist aesthetics is part of the wider distrust of mechanistic thinking in deep ecology and New Age ecophilosophy. A commonplace view in so-called 'new paradigm' ecophilosophy is that mechanistic methods of knowing the world, which try to understand phenomena by reducing them to their component parts, are always inferior to the more holistic view taken by advocates of organicism. This is a central thesis in the critique of reductionism in modern science by Theodore Roszak, Fritjof Capra, Carolyn Merchant and Morris Berman, amongst many others, and is a central principle of ecopsychology and ecophenomenology (Capra 1992; Berman 1981; Merchant 1982).

These radical environmentalists tend to value organicism more than mechanism, in a way that has important implications for the study of music. Yet, as Raymond Williams usefully pointed out, in sixteenth-century Europe the two terms were synonymous; they became opposites in the nineteenth century, reflecting the recent division of the natural sciences into biology, which was concerned with living processes, and physics, which studied mechanical ones (1983: 228). Moreover, the rejection of mechanism by many radical environmentalists is based on a caricature of scientific methodology. A more nuanced account of science reveals that both organicist and mechanistic models can be equally valid in different methodological contexts. As analytical philosopher Robert Kirkman comments, radical ecophilosophers tend to support organicism 'not because they have achieved some sort of privileged insight into the nature of things but because they are convinced of the vital political necessity of opposing mechanism and its consequences. As a consequence, environmental philosophers tend to work backward from the conclusion they would like to support to the first principles that strike them as best suited to the task of supporting those conclusions' (2002: 44).

Politically, the main problem with naturalistic aesthetics is their tendency towards 'naturalisation'; that is, presenting a particular state of affairs as 'natural', and therefore inevitable and unchangeable, even though there are many competing ideas and uncertainties about what 'nature' actually is. Claims about naturalness and authenticity can obscure the social constructedness, and therefore changeability, of that state of affairs. As ecocritic Mei Mei Evans points out, the politics of naturalisation tend to work against socially subordinated groups in society, such as people of colour, women and gays, whose behaviour is often judged negatively against dubious notions of what is considered 'natural' (Evans 2002: 181-93).

Naturalisation, then, is a process whereby people use organicist rhetoric in an attempt to pass off their own political ideology as universal truth. Historian Anna Bramwell has shown that the concept of 'nature' has been called upon to justify a wide range of political positions, both Right and Left, as the nature-worshipping National Socialists in the 1930s tragically demonstrated (Bramwell 1989). But social and political ideas, wherever they lie on the political spectrum, cannot be derived directly and unproblematically from a study of the natural world. Ecological science, as Robyn Eckersley observes, 'does not tell us why we *ought* to orient ourselves toward the world in a particular way. It can inform, inspire, and redirect our ethical and political theorising, but it cannot justify it. *That* is the task of ethical and political theory' (1992: 59). Human beings are relatively free to choose their ethical, political and aesthetic responses to ecological problems, which may be guided by current ideas about how the natural world works, but should not seek justification from them. The value of the concept of the ecological self is political, in this sense, and does not depend for its significance on its supposed basis in nature.

A qualification should, however, be added to this critique of naturalisation, in that some anthropologists see it as a source of the very social cohesiveness and sense of collective identity in indigenous societies which is markedly lacking in their own. Steven Feld writes that, for the Kaluli people of Papua New Guinea, music plays a central role in fostering what he calls a 'felt naturalness of the whole, as one finds oneself in and through the music and the music in and through oneself' (1994: 122-3). He quotes anthropologist Howard Becker: 'people do not experience their aesthetic beliefs as merely arbitrary and conventional; they feel that they are natural, proper and moral'. From this perspective, wholesale rejection of naturalisation is a symptom of Western alienation from nature. Even if there is no way back from this alienated state to an unmediated nature, then, it is nevertheless important to recognise that just because authenticity is culturally and historically relative does not mean that it is not experienced as real.

## 3.2 American popular music and the pastoral mode

The favoured modes of environmentally concerned popular music have been elegy and satire, because both explore what ecocritic Jonathan Bate calls 'the contradiction between actuality and the ideal' (2000: 73). The ideal tends to be a pastoral landscape in which human beings feel at home in the natural world; elegy is a lyrical meditation on its loss, and satire a denunciation of those deemed responsible for that loss.

The pastoral, then, is the main mode by which the eco-utopian potential of music has been articulated in American popular music since 1960. Indeed, the pastoral mode has come to dominate American musical representations not only of the rural, 'middle' landscape, but also of wilderness. As ecocritic Andrew Light notes, according to the classical view prevalent in the Puritan era, wilderness was hostile and evil, and its conquest a sign of social progress. Wilderness spaces were marked as separate from human civilisation, and their inhabitants accordingly demonised as nonhuman beasts. Civilisation and its inhabitants, in contrast, were celebrated as superior to wilderness (1999: 139).

This classical notion of wilderness began to change in the nineteenth century, when landscape aesthetics came to be influenced by what art historian Barbara Novak calls a 'Christianised sublime', according to which the wilderness became sentimental rather than awesome, suitably tamed for the growing number of middle-class tourists in the American West (1995: 44). By the twentieth century, what environmental historian William Cronon calls the 'domesticated sublime' became the preferred mode for mass media constructions of wild nature (1996: 75). In American musical culture, the domesticated sublime is a useful description of many musical depictions of wilderness, from the country-pop of John Denver discussed in Chapter 5, to the jazz of Paul Winter in Chapter 12. In the 1970s, as we will see in Chapter 9 on the rock music of that decade, the endangered wilderness came to be symbolised in popular song by two recurrent tropes: the whale and the rainforest.

For Ernst Bloch, the pastoral mode in Western classical music embodied utopian hopes for a humanised natural world achievable through an apparently benign form of industrial progress. He argued that when Beethoven depicted a storm in his Symphony No. 6, the 'Pastoral' (1808), he moved Romantic nature music beyond simple tone painting of the surface effects of the natural world, that is of 'natura naturata', towards a more profound revelation of its inner workings, or 'natura naturans'. The pastoral tradition after Beethoven went on to reveal what a properly humanised nature would feel like. Bloch thus praised the musical depiction of bleating sheep in Richard Strauss' *Don Quixote* (1897) and of an Alpine landscape in the first movement of Gustav Mahler's Sixth Symphony (1904), in which, he wrote, the composer 'allows the voices of nature to break in, always with a ray of hope or the light of a saviour' (Bloch 1995: 1083). For Bloch, then, the utopian promise of the pastoral mode lay in its promise of transcending the limitations of nature through technological transformation.

In *Aesthetic Theory*, Adorno took a more ambivalent attitude to the pastoral mode. He argued that nature imagery serves two ideological functions in capitalist societies, one conservative and the other radical. In its conservative usage, nature imagery is an allegorical defence of the status quo, because it suggests that people are already happily reconciled with nature in the present, capitalist society. As he put it, if this allegory 'were falsely claimed to be a state of real reconciliation, then natural beauty would be reduced to a means for justifying and disguising antagonisms in a society in which such natural beauty is still possible'. The more progressive use of nature imagery, in contrast, draws on its utopian potential for radical social change. Here, to turn nature into an aesthetic object is to posit its beauty as an allegory of 'the possibility of a sphere beyond bourgeois work and commodity relations' (Adorno 1984: 102).

Adorno's dialectical use of the pastoral resembles discussions in ecocriticism, which themselves can be reapplied from literary criticism to music. The idea that conservative uses of nature imagery are an escapist alibi for the status quo is similar to Leo Marx's category of the 'simple pastoral' in *The Machine in the Garden*. Marx's distinction between two kinds of pastoralism, the 'popular and sentimental' and the 'imaginative and complex', was based on a firm opposition between low and high culture (Marx 1964: 5). 'The mass media', he wrote, 'cater to a mawkish taste for retreat into the primitive or rural felicity exemplified by TV westerns and Norman Rockwell magazine covers' (6). Such artefacts are a source of 'infantile wish-fulfilment dreams, a diffuse nostalgia, and a naive, anarchic primitivism' (11). Recalling Adorno, then, Marx argued that the simple pastoral is pernicious, in that it has 'appeared with increasing frequency in the service of a reactionary or false ideology, thereby helping to mask the real problems of an industrial civilisation' (6).

Although he acknowledged that the 'yearning for a simpler, more harmonious style of life, an existence "closer to nature" is the psychic root of all pastoralism – genuine and spurious', Marx proffers works of 'high' literary culture as 'genuine' pastoral, because they put the same impulses as the 'popular' pastoral to different ends (6). Writers such as Nathaniel Hawthorne and Henry David Thoreau created a 'complex' pastoral when they recognised the forces of industrialisation as a threat, or what Marx called a 'counterforce', to the pastoral idyll. In contrast to the simple pastoral, then, the complex pastoral creates a sense, not of simple repose, but of 'dislocation, conflict, anxiety'. Such works, wrote Marx, 'manage to qualify, or call into question, or bring irony to bear against the illusion of peace and harmony in a green pasture' (25).

In its intellectual and political sophistication, the complex pastoral resembles Bloch's notion of concrete utopia, whereas the simple pastoral serves the same function as his abstract utopia. Both pairs of notions will be important in this book. However, we will see how the complex pastoral is not solely restricted to 'high' culture, as Leo Marx suggested, in that irony and qualification can also be seen in the folk, rock and country songs of the period since 1960, in the work of artists as diverse as Pete Seeger, Bob Dylan and Captain Beefheart.

In *The Environmental Imagination*, Lawrence Buell follows Leo Marx in attempting to revive the critical reputation of the pastoral mode, by similarly arguing that it can be more than simply a conservative retreat from modernity that acts as a safety valve for dominant political interests. The pastoral can also take on reformist or even revolutionary aspects, particularly when the act of valuing beautiful landscapes implies opposition to industrial development. 'American pastoral representation cannot be pinned to a single ideological position', Buell writes. 'Even at its seemingly most culpable – the moment of wilful retreat from social and political responsibility – it may be more strategised than mystified' (Buell 1995: 44). He cites as literary examples Rachel Carson's *Silent Spring*, and Wendell Berry's *The Unsettling of America* (1977), both of which used pastoral imagery to mount their critiques of industrial agriculture.

The critical potential of the pastoral mode is an important concept in the arguments about music and nature in this book. Yet the pastoral has its problems as a conceptual model for political ecology, particularly if it implies an anti-urbanism that is of questionable value for environmental politics. Indeed, such anti-urbanism contradicts the urban origins of the pastoral mode, which was invented by urban freemen in ancient Greece, and has always been a sign of the separation from agrarian life of the urban, literate societies that have embraced it (Gifford 1999: 15). As musicologist Philip Tagg notes, this is equally true of the American musical scene which inherited the pastoral mode from Western Europe. He writes that 'the development of a class society and a clear division between town and country seems to be necessary historical precedents for treating nature as a distanced aesthetic object in music' (1982: 4). Moreover, like the pastoral mode, American popular music itself is an urban product. In the United States, as elsewhere in the world, the city has been a vital meeting place for what the French humanist Marxist Henri Lefèbvre called forms of 'assembly, encounter and simultaneity', and therefore has been the main site for the production and consumption of music, even music such as ‘country’ which is purportedly about the countryside (1991: 149).

Despite its urban origins, then, the corollary of the pastoral ideal is that cities are artificial and unnatural. Geographer David Harvey writes that it is ‘almost as if a fetishistic conception of “nature” as something to be valued and worshipped separate from human action blinds a whole political movement to the qualities of the actual living environments in which the majority of humanity will soon live’ (1996: 427). Yet the city can be easily defended from a humanist perspective, for the individual freedom it permits. Big cities also allow for economies of scale when it comes to providing infrastructure such as water, electricity and schools. But ecological arguments in favour of cities are more complex and uncertain, because cities leave big ‘ecological footprints’. Political scientist Peter J. Stoett asks: ‘If cities produce more pollution than can be safely managed, if traffic congestion not only creates smog but actually impedes transportation, if neighbouring land is degraded because of waste disposal needs and excessive demands on natural resources, and if most rural-urban migrants end up in unsanitary and powerless shanty towns, is urbanisation really a step in the right developmental direction?’ (1994: 342). Whatever the ecological value of cities, the political, economic and technological question of whether cities can be transformed into ecologically benign spaces tends to be suppressed in the pastoral’s celebration of ‘middle’ or agrarian landscapes as ideal spaces for human habitation.

The anti-urbanism inherent in the pastoral mode is matched by the class, racial and geographical divisions it often masks. As recent environmental historians and ecocritics have shown, the cult of the pastoral does not have universal appeal. Instead, it remains an ideal mainly for the white, urban middle-classes. The class-based elitism of the pastoral has been noted by historian Anna Bramwell, who points out that pastoral contemplation is caught up in what economists call the problem of the ‘positional good’; that is, the conundrum produced by rising standards of living whereby ‘some goods could only be enjoyed if they were confined to a few; cars and motor travel were the classic examples’ (1989: 215). As Buell notes, the pastoral emphasises ‘alone-with-nature experiences’ rather than ‘communal betterment’, and is therefore perfectly compatible with the possessive individualism on which the values of economic liberalism, and therefore capitalist development, are based (2001: 38).

Moreover, for both African-Americans and the white working class, the rural landscape has historically been associated with hardship and suffering, rather than pastoral contemplation. Slavery, as ecocritic Michael Bennett puts it, 'created a system whereby those of European descent controlled a pastoral landscape that included those of African descent as part of their property' (2001: 205). The pastoral mode hides this racial history of violence and exploitation, and relegates black people to its idealised scenery. Despite this, as we will see in Chapters 9 on hip hop and 12 on jazz, the pastoral mode has been an important means by which African-American musicians have expressed their hopes and fantasies about a more satisfying, and less destructive, relationship between human beings and the natural world. Buell describes the pastoral tradition within the literature of the African diaspora which evokes 'a traditional, holistic, nonmetropolitian, nature-attuned myth of Africanity in reaction to and critique of a more urbanised, "artificial" European order – and evokes it, furthermore, from the standpoint of one who has experienced exile and wishes to return' (1995: 64). Regarding African-American music, Cecelia Tichi observes that both 'Negro' spirituals and Southern gospel music inherited from Christianity 'a symbolic sense of space in which the Promised Land is a pastoral release from earthly suffering' (1994: 186).

Yet an anti-pastoral tradition has also arisen in American popular music to articulate a sense of alienation from nature. The song 'Strange Fruit', written in 1937 by the white Jewish union activist Abel Meeropol under the pen name of Lewis Allan, debunked the 'pastoral scene of the gallant South' to reveal the 'bulging eyes' and 'twisted mouth' of racial violence; mingling with the scent of magnolias is 'the sudden smell of burning flesh'. Blues lyrics have also used the anti-pastoral to comment on the harsh realities of the American environment, whether their cause is human and political, as in 'Strange Fruit', or natural and beyond human control, as in the Masked Marvel's 'Mississippi Boweavil Blues', collected in Harry Smith's *Anthology of American Folk Music* (1952), or accounts of the 1927 Mississippi flood in Memphis Minnie's 'When the Levee Breaks' and John Lee Hooker's 'Tupelo'. Country music, as we will see in Chapter 5, also has a strong tradition of anti-pastoral, which has the critical potential to expose the harsh realities of human relationships with the natural world which the pastoral mode merely evades. The aesthetically pluralist approach adopted in this book is open to the artistic potential of both the pastoral and anti-pastoral modes.

The idea that music has utopian potential to foster ecological awareness raises a crucial question: how does this process work for the listener? Various musicians and philosophers have speculated on what might be called 'eco-listening'. In doing so, they establish more important concepts for a study of ecocriticism and popular music, which will be explored in the next chapter.

# Chapter Four

# Eco-listening

A recurrent claim made by composers and musicians influenced by both deep ecology and New Age ecophilosophy is that listening to music has the potential to be an ecologically attuned activity. The notion that music is the art form best suited to fostering the ecological self is a corollary of the long-standing suspicion of visuality in Western culture, which holds that the dominance of the visual sense in human beings encourages a sense of separation between subject and object, or human perceiver and things in the world, which has had disastrous consequences for the health of the environment. 'Vision', writes the Canadian environmental philosopher Neil Evernden, 'permits us the luxurious delusion of being neutral observers with the ability to manipulate a distant environment' (1985: 84).

For some ecophilosophers, the sense of hearing overcomes the limitations of sight by enacting the fundamental ecological principle of holistic interconnectedness. Drawing on Martin Heidegger's opposition to modern technology, the British ecopsychologist David Kidner writes that the 'technological world's […] overreliance on visual perception is an initial step in the process of destruction, fragmenting the world in a way that makes its subsequent physical fragmentation seem quite unremarkable and leading us toward a world populated only by "things"' (2000: 305). In opposition to this tendency, Kidner celebrates the auditory cultures of indigenous peoples, who, he asserts, have a more ecologically sensitive relationship with their environment than modern Westerners.

The most far-reaching claims of this type involve the notion of 'immersive listening', which, for some musicians and cultural critics alike, is a means of attaining a new, apparently enlightened state of consciousness in which the individual self merges with the rest of the natural world. A more scientific notion of musical effects is that of 'entrainment', which has also been linked in New Age thought to ecological awareness. This chapter describes and evaluates both of these theories of what may be called ‘eco-listening’.

## 4.1 Immersive listening and the deep ecological self

Theodor Adorno made a firm distinction between popular and art music based on the different modes of listening they encourage. Whereas art music deploys unfamiliar forms to challenge the listener, popular music merely requires what he called ‘regressive’ listening appropriate for the passive consumption of commodities. Regressive listening, he wrote, appears ‘as soon as nothing is left for the consciousness but to capitulate before the superior power of the advertised stuff and purchase spiritual peace by making the imposed goods literally its own thing’ (Adorno 2002: 304). In order to resist the manipulations of the capitalist music industry, Adorno advocated that listening to music should be a conscious act of mental concentration. Music, he wrote, has always been an attempt ‘to transform the indolence, dreaminess, and dullness of the ear into a matter of concentration, effort, and serious work’ (Eisler and Adorno, 2004: 75). For him, the difficult, atonal modernism of Arnold Schoenberg exemplified music that demanded such attentive listening. Only through mental work of this kind, he believed, could the listener begin to nurture the social and political awareness that, for Adorno, was the true ethical significance of art.

The notion of ‘immersive’ listening is in marked contrast to Adorno’s modernist aesthetic, and may be seen as part of the post-modern turn in music aesthetics explored in Chapters 1 and 2. In his book *Ocean of Sound*, British musician and composer David Toop traces the image of immersion through many different types of music, from the French impressionism of Debussy and Ravel, through mini-malism, to popular musics such as rock, dub reggae and techno. He notes that a wide range of musical techniques can produce immersive sound, including deep bass, echo, dense orchestration and noise (in both senses of the word: high volume and unpitched sound). The link between immersive listening and ecological thought focuses on the idea of music as a vibration that works on the unconscious or intuitive parts of the human mind-brain. When the volume is high, writes Toop, music is ‘felt at its vibrational level, permeating every cell, shaking every bo-ne, derailing the conscious, analytical mind’ (1995: 273). In the 1960s, rock, jazz and minimalism were all seen by some aficionados as means of attaining what Eastern mystics call 'cosmic consciousness', or 'the One', a state of supposedly expanded awareness gained through instinct and feeling rather than logic and reason. Used in tandem with music, the new mass technology of LSD, argued its leading advocate Timothy Leary, could also gave instant access to the One, without the need to undergo the more rigorous and demanding disciplines of Eastern or Native American religious practices (Leary 1968).

In the rock music of the sixties, the *musique concrète* of the Jefferson Airplane's 'Crushingura' (1968) and the feedback on The Grateful Dead's album *Live/Dead* (1969) both explored noise as unpitched sound. Yet it was the sheer amplified loudness of rock, rather than its usually more conservative attitude to pitch and tonality, that was central to its quasi-mystical promise. Electricity has always attracted mystical thinking, because it is an invisible force that acts at a distance from its object, like magic. In the 1960s, American rock music literally amplified and democratised this Romantic cult of electro-mysticism whose classic expression, at the start of the twentieth century, was Henry Adams's comparison of the 'silent and infinite force' of the electrical Dynamos at the 1900 Great Exhibition in Paris with statues of the Virgin Mary (Adams 1961: 379-90).

In his 1970 article 'Rock as Salvation', Benjamin DeMott celebrated the new rock music as a new electronic means of mass consciousness transformation. 'The rock experience at its most intense', he wrote, 'is an imitation of engulfment and merger, a route to a flowing, ego-transcending oneness. As fans and enemies alike know, rock sound overwhelms separateness, the mental operations that discern and define here and there, me and not me [...] Pounded by volume, riddled by light, the listener slides free from the restraining self and from the pretences of a private 'unique' rationality' (DeMott 1970: 199-200). DeMott here recalls Theodore Roszak's proto-ecological critique of the American military-industrial complex in *The Making of the Counter-culture*, which similarly endorses a holistic and monistic theory of the self (Roszak 1970).

In writings on music, the notion of immersive listening as egoless merger with the Other is often gendered as feminine. In *The Sex Revolts*, Simon Reynolds and Julie Press begin to bring out its eco-feminist implications. Although masculine rebelliousness in rock music is usually solipsistic and narcissistic, they write, some male rebels 'tire of the sterility of individualism, and move towards affirming the transindividual – from insularity to oceanic feelings. Born-to-run nomadism succumbs to a longing to return to the maternal bosom, a sanctuary that can take the form of an idealised woman, or a mystical investment in Mother Nature and the Cosmos' (Reynolds and Press 1995: 156).

Before exploring the environmentalist aspects of such claims, a brief note on the psychoanalytic interpretation of music is necessary. The above reference to 'oceanic' fantasies invokes Freud, who argued that the 'feeling of an indissoluble bond, of being one with the external world as a whole' was an indication of 'primary narcissism' (Freud 1963: 2). For Freud, such fantasies recall the infant's early experience of plenitude with the mother, a feeling lost when the child enters the Oedipal stage of development, and learns to be a separate individual. For Freud, then, oceanic fantasies were a regressive and immature form of neurosis.

In the feminist psychoanalytical theories of Julia Kristeva, in contrast, pre-Oedipal fantasies are viewed more positively, as subversive of what she sees as the patriarchal codes of rationality imposed on the child by its entry into the 'Symbolic Order' of language and culture. What Kristeva calls the pre-Oedipal 'chora' is a state of undifferentiated, somatic bliss which the child experiences before he or she is forced by society to acquire both language and an individual identity (Kristeva 1984: 25). From this radical psychoanalytic perspective, music is a 'sonorous envelope' that recalls the voice of the mother enclosing her child in protective bliss (Silverman 1988; Shepherd and Wicke 1997: 59-87). According to this theory, then, the immersive nature of music is politically subversive, in that it threatens the illusion of a separate, individual ego on which bourgeois society is founded.

Although such psychoanalytical theories of immersive listening rightly attempt to account for music as a non-linguistic and somatic phenomenon, as musicologist Robert Fink suggests, they risk becoming too reductive, locking the critic into a single explanatory model which views music solely as a psychic construct linked to childhood, rather than a product of particular historical and material formations (2005: 6). Bearing these reservations in minds, it is with the latter approach that this book is more concerned.

The ecological implications of immersive listening, nascent in Benjamin DeMott's notion of rock music as salvation, were revived in 1989 by William Irwin Thompson, the New Age cultural historian and founder of the Lindisfarne Association. His essay 'Pacific Shift' links immersive listening with questions of ecology, technology and politics. Thompson revives the countercultural utopianism of the 1960s by celebrating the immersive loudness of rock music as indicative of a 'new electronic culture-ecology'. The teenagers who go to rock concerts, he writes, 'understand that noise is the new form of identification with the group'. If they are risking deafness, he adds, this is the price they pay for being part of the new future collectivity: 'If you are willing to lose the membrane of your biological integrity to play with a group, or join your body to the new body politic in a rock concert then what you are demonically signalling is the emergence of the next level of historical order, the shift from literate civilisation to … what shall we call them? Noetic polities? Noetic plasmas?' (Thompson 1989: 35). Thompson's reference is to the Jesuit mystic Teilhard de Chardin, who believed that the next stage in human evolution will culminate in a single, cosmic mind formed out of the collective mass of human intelligences (Capra 1982: 331-2). Echoing Marshall McLuhan (1967), Thompson enthusiastically announces the new electronic culture-ecology as a shift from a hierarchical world dominated by the industrial nation-state to a newly democratised 'planetary culture.'

The optimism and totalising sweep of such writing is seductive. However, there are problems, both musicological and philosophical, with Thompson's ecophilosophical interpretation of rock music as immersive listening. From a musicological perspective, as Theodore Gracyk argues, the haptic qualities of loud rock music, the fact that it is felt as much as heard, may actually reinforce an atomistic sense of self, rather than break it down. As he puts it, 'if touch is basic to maintaining a line between self and nonself, the typical volumes of rock music may only reinforce a passive and atomistic sense of the self. Too often, rock's volume functions as a filter on the world, fostering withdrawal rather than facilitating communication' (Gracyk 1996: 115). Gracyk's argument applies equally to other forms of loud, bass-driven popular music, such as techno, house or reggae. Evan Eisenberg observes that the rock music constructs a self that is 'at once eagerly social and deeply solipsistic – a condition of adolescence' (1987: 53).

In any case, for its fans, the high volume of rock music usually signifies modernist alienation from society, and is indicative of a politics of symbolic insurgency and *épater la bourgeoisie* rather than of integration with a benign natural world. In his essay, 'A Reasonable Guide to Horrible Noise', the American rock journalist Lester Bangs wrote that for those who believe that 'the life force is best represented by the livid twitching of one tortured nerve, or even a full-scale anxiety attack', then 'the shriek, caterwaul, the chainsaw gnarlgnashing, the yowl and the whiz that decapitates may be reheard by the adventurous or emotionally damaged as mellifluous bursts of unarguable affirmation' (1987: 301). Sharing this confrontational relish for shock and nihilistic despair is Trent Reznor of Nine Inch Nails, the industrial noise band from Cleveland, Ohio, who asked the musical question: 'I can make something loud, but how can I make it the loudest, most abrasive thing I've ever heard?' (Cold 1994: 53). This modernist pleasure in the aural violence of excessive noise recalls not the benign nature constructed by some radical ecophilosophers, but rather the sublime nature aesthetics described by the eighteenth century British philosopher Edmund Burke: 'excessive loudness alone is sufficient to overpower the soul, to suspend its action, and fill it with terror' (Nye 1994: 246).

The monist idea of the self immersing itself in a benign universe through music is equally problematic from the point-of-view of ecophilosophy. Australian ecophilosopher Val Plumwood wrote of the dangers of sinking into what she called an 'Ocean of Continuity', in which the self becomes indistinguishable from the rest of the universe. She observes that such an abstract identification with the cosmos is unlikely to provide a firm basis for an ethics of care about one's environment, which is more likely to emerge from social and personal ties to specific places and landscapes. As she puts it, the 'analysis of humans as metaphysically unified with the cosmic whole will be equally true whatever relation humans stand in with nature – the situation of exploitation of nature exemplifies such metaphysical unity equally as well as a conserver situation, and the human self is just as indistinguishable from the bulldozer and Coca Cola bottle as the rocks or the rainforest'. Plumwood proposes instead an alternative ethical model that recognises 'not only our human continuity with the natural world but also its distinctness and independence from us and the distinctness of the needs of things in nature from ours' (Plumwood 1993: 177-8). Monistic self-merger is therefore not the only alternative to the egoism on which the liberal capitalist self is based.

## 4.2 Rhythmical entrainment and New Age ecologism

Ecological arguments about immersive listening extrapolate from the more widely acknowledged role that music plays in building a sense of communal identity. Music has indeed played a vital role in environmental protest groups all over the world, helping to foster shared belief systems and to promote ideological and group unity. As singer and environmental activist Darryl Cherney commented, because Earth First! is 'essentially a warrior tribe, and we're dealing with [... ] insurmountable, or seemingly insurmountable obstacles, we have a special need for the release of tension and sorrow, as well as a need to express our joy in the good fight' (Lee 1995: 49-50). Building on this recognition of music's role in the formation of collective identity, theorists within the deep ecology and New Age movements have begun to speculate on the role that it can play in fostering a sense of 'community' that is expanded to include not only human beings but also the natural world itself.

One of the most influential statements of the notion of ecological community was made by the American conservationist Aldo Leopold, whose 'land ethic [...] simply enlarges the boundaries of the community to include soils, waters, plants, and animals, or collectively: the land' (1948: 204). While Leopold's call for an extension of ethical consideration to the natural world remains a crucial insight for environmental politics, its basis in science has been questioned by neo-Darwinists, who reject the idea that the natural world really is a 'community' of living organisms; instead, they understand habitats as characterised by competition, either between species, between individuals within those species, or between individual genes. Biologist Stephen Jay Gould argued that 'notions of community and natural harmony, however illuminating as metaphors, do not reflect nature's primary evolutionary unit, the population of individuals within species' (1990: 183). Many ecologists, as ecocritic Dana Phillips summarises, 'now see concepts like cooperation and social organisation, when applied to the natural world, as ambiguous at best and irrelevant at worst' (2003: 49). He notes that the most productive areas of ecological science have been population ecology and conservation biology, both of which take a neo-Darwinian perspective. When radical ecophilosophers assert that the natural world is a 'community', then, they tend to be making a political rather than a scientific point. In doing so, they risk conflating political arguments with scientific ones in order to give authority to their claims.

Bearing these reservations in mind, we will now examine in detail the New Age conception of music as ecological community. The British composer June Boyce-Tillman, whose ideas we first approached in Chapter 2, has developed her interest in music therapy into both a vision and a program for social and political change, based on an ecocentric theory of music as a force that reconnects humans with other living beings in an expanded, cosmic community. Her ecological vision is, in effect, a rediscovery of ancient religious attitudes to music. In ancient Greek culture, she notes, Hermes was the god of music, medicine and theology. Pagan societies saw music as a way of communicating with the gods whom they believed inhabited an enchanted natural world. In these societies, writes Boyce-Tillman, 'people sang to apple trees so they would increase their yield and animals were encouraged to give up milk with songs' (2000: 39).

This benign state of affairs began to change, however, with the establishment of Christianity in the European Middle Ages, which saw a decline in such religious conceptions of music, as the new state religion emphasised 'domination over nature' and the 'humanity of God'. By the time of the Enlightenment, the idea of the cosmos as a natural community had declined further in the face of humanist and anthropocentric views of the natural world. Even in the nineteenth-century Romantic cult of divine nature, observes Boyce-Tillman, the 'flow' between human beings and the natural world was conceived as a one-way process, within which human beings could be inspired by nature, but not vice-versa. There was, she writes, 'no notion that the natural world could in turn be affected by music. Ideas that the material world could be influenced by sound disappeared post-Enlightenment as scientific realism gained sway'. Her optimistic conclusion is worth quoting at length:

> It has resurfaced in the New Age. It has rediscovered ancient notions that music is vibration and that the whole of the natural world is in vibration. In this system it is not seen as surprising that music will affect the natural world. The walls of Jericho could have fallen if Joshua found the right note, and sound can shatter glass. In concentrating on the destructive potential of music, the possibility of using music to create and build in the natural world has deliberately been ignored or ridiculed. This is counter to the dominant culture. Music and healing in the New Age are now being related in a cosmic context. Humans are healed by being restored to a

> relationship not only with other humans but to relationship with the natural world (39).

The rhapsodic prose style is as enticing as William Irwin Thompson's, whose writings were quoted earlier in this chapter. Yet Boyce-Tillman's ecopsychological theory of music is similarly based on questionable assumptions, most obviously that magic has real effects. By preferring mystical views of the world to 'scientific realism', she overstates music's powers of transformation. Musicologist Victor Zuckerkandl warned against such excessive irrationalism when accounting for musical effects. As he put it, 'those who have lost hope in the word believe they hear in music the glad tidings of ultimate return to a lost paradise'. Such 'overstatements', he continued, are made possible by 'relegating music to the status of magic' (Zuckerkandl 1973: 73). Indeed, Boyce-Tillman's advocacy of music as a form of benign magic can be placed within wider debates about the politics of the New Age movement as a whole. As Andrew Ross argues, there is a tendency for New Agers to prefer mystical solutions to the more materialist, political project of social ecology. 'Without any alternative (left) politics of social individualism', he writes, 'it is no surprise to find that people will opt for spiritualist solutions for the problems encountered in a materialistic culture' (1991: 68-9).

Perhaps the most influential source for mystical notions of music and nature is *Sounding the Inner Landscape* (1990), by the New Age composer and 'music healer' Kay Gardner. Drawing on an Aruveydic model of the body, Gardner asserted that each musical key has its own specific healing effect on the body's *chakras*, or energy fields (1990: 26-7). Music is thus a form of cosmic energy that can act as a healing force in and of itself. Gardner's influence can be seen not only on Boyce-Tillman, but also on the American folksinger and environmental activist Alice Di Micele. In a 2002 interview in the New Age magazine *Magical Blend*, she mentions her interest in Gardner's research on 'which *chakras* get activated by which note', and goes on to talk about how the 'act of music' is a 'healing force' and a type of 'magic' (Dobra 2002: 55-7).

For Di Micele, the opportunity for collective singing offered by folk performance can provide the audience with an immersive experience of planetary ecological healing. She tells of a 'sounding' ceremony in which she participated with the white witch Zsuzsanna Budapest, who explained that 'when our bodies are vibrating together and making a tone, that creates an energy and a resonance, and that can be focused into magic, if you have that vibration with intention'. Di Micele adds that, when she performs, she tries 'to get people to sing, and when they're singing the lyrics, there's intention in the words, and so I'm praying that we're creating the vibration together for healing the planet'. Folk music, she believes, can encourage a holistic merging that transcends the 'illusions of religion or colour or class or status or sexual orientation' (57). Unlike speech, the vibrational quality of music 'actually has the ability to transform us on a cellular level. It's moving energy' (56). For Alice Di Micele, then, as for Boyce-Tillman and the jazz musician Kali Fasteau, whose work will be discussed in Chapter 12, music's ontological status as cosmic energy is the key to its power in fostering a sense of communion with the natural world.

These New Age conceptions of music as cosmic vibration conflate the ancient Aruveydic notion of energy fields with the scientific theory of 'entrainment', or the mutual phase-locking of two or more oscillating bodies. This phenomenon was first discovered in 1665 by the Dutch scientist Christian Huygens, who observed that two pendulum clocks placed next to each other eventually develop a shared beat. Scientists have used the theory of entrainment to account for the physiological changes that take place in both musicians and audiences alike when engaged in music-making (Clayton et al 2005: 3-75). Anthropologist Judith Becker defines rhythmical entrainment as the phenomenon according to which our bodies and brains 'synchronise gestures, muscle actions, breathing, and brain waves while enveloped in musicking' (2001: 151). She notes that it produces a transformation of consciousness that is both individual and collective, and is responsible for the feeling of well-being people often experience when music-making is successful.

The concept of entrainment has become important in music therapy, where the technique of Rhythmic Entrainment Intervention is used to effect changes in the client's nervous system (Ortiz 1997: 317-325). It has also become a central concept for some composers and musicians, in world music, jazz and electronic dance subcultures, to articulate their understanding of the relationship between music and ecological thinking, as we will see later in this book.

The scientific study of entrainment combines approaches from contemporary neuropsychology and evolutionary musicology. Since the development of Magnetic Resonance Imaging in the 1980s, neuropsychology has made great advances in the study of the physiological effects of music on the human brain. Such work is complemented by the more speculative field of evolutionary musicology, which explores the possible role of music in human evolution. This work has produced two main explanations for music as an adaptive function: to attract sexual partners (Darwin's own position), and to encourage group bonding. Neuropsychologist Isabelle Peretz writes that music 'possesses two design features that reflect an intrinsic role in communion (as opposed to communication, which is the key function of speech). Pitch intervals allow harmonious voice blending when sounding together, and temporal regularity facilitates motor synchronicity. These two musical features are highly effective in promoting simultaneous singing and dancing' (2001: 115). The phenomenon of motor synchronicity to which Peretz refers is another way of describing rhythmic entrainment, which, according to this theory, is a means by which music may work to 'override "selfish genes" for the benefit of the group'.

Although there is always a danger that evolutionary musicology, like the evolutionary psychology on which it draws, can become simplistic and reductive, scientific research of this kind is at least attempting to understand the appeal and effects of music without recourse to religious faith or New Age mysticism. As far as the theories of music and ecology explored in this chapter are concerned, we may conclude that any powers of physical, mental or social transformation which music may have will result from a combination of social, psychological and physical factors. That music does not bring about social or personal changes in itself was a lesson learned by many in the 1960s counterculture. In 1968, Ed Sanders, poet, singer and composer with The Fugs, sang a much paraphrased sentence from Plato's *Republic*: 'When the mode of the music changes / the Walls of the city shake' (1974: 191). In the late 1980s, reviewing the excessively optimistic view of music he held in his youth, Sanders commented: 'I thought that if you yodelled the right yodel, it was like a mystic mantra that changed the world' (Denselow 1989: 84). This near-mystical power attributed to rock music in the 1960s ultimately contributed to the twin delusions of idealism (the idea that consciousness can change the world on its own) and voluntarism (that the world can be changed by individual acts of will alone), both of which severely limited the political effectiveness of the counterculture. As Eric Mottram put it, the sixties counterculture believed that revolution would be effected 'through change in the soul, through mind-blowing sounds and words' (1988: 190). Music, he continued, 'was supposed to be the supreme power to counter the mercenaries, and had been built up to absurd heights of political capacity' (212). The danger is that the New Age ecophilosophies examined in this chapter are falling into the same trap again.

Listening to music, then, is an activity determined by biological, cognitive and social factors in a more complex way than that suggested by New Age and deep ecological philosophies. Moreover, such theories of music tend to assume a transcendental, universal human subject, often constructed in the third person plural ('we'). However, despite the fact that the basic physiology of human hearing is a cross-cultural universal, listening itself, as a cognitive activity, is socially relative. Philosopher John Andrew Fisher observes that our aesthetic responses to particular sounds are underdetermined by the characteristics of the sounds themselves; that is, they are only guided by the sounds, rather than totally determined by them. The meaning of a particular sound is therefore context-dependent, so that there can be many different ways of listening to a given sound (1998: 172). Musicologists Shepherd and Wicke similarly point out that the sounds of music do not 'somehow contain within them meanings which are then instilled or articulated in listeners in a direct and mechanical fashion' (1997: 115). In this light, for the act of listening to music to lead to greater integration with one's environment will depend on the listener's pre-existing knowledge of, and support for, the theory of the ecological, relational self on which the theory depends.

Moreover, these ecophilosophical claims ultimately fall into the same trap as the theories of visuality from which they emerge: that of isolating one sense from the others, while ignoring the ways in which, in everyday human perception, all of the different senses interact. Indeed, Henri Lefèbvre notes that odours play an important role in denoting what he calls 'nature's violence and largesse' (1991: 198). Greater environmental awareness may emerge, then, not from encouraging our sense of hearing alone, but by fostering a better balance of all the senses (Rodaway 1994). But even this enhanced sensory awareness, which can certainly be encouraged by an appreciation of music, and of the arts in general, will only be a small step towards positive environmental change, which will depend much more on questions of politics, economics and history too often omitted from the kind of ecophilosophical discussions examined in this chapter.

# Part 2

# Ecocriticism and American popular music since 1960

# Chapter Five

# Blues and country music

Musicologist Philip Tagg notes that the musical features that musicologists often cite as characteristically 'black' or 'African-American', such as 'blue' notes, call and response, syncopation and improvisation, all exist in other forms of music originating in all parts of the world, including Europe, and, conversely, do not appear in all musics played by black people in all parts of the world (Tagg 1989). These racialised terms, often taken for granted in discussions of popular music, are thus historical constructions that change over time, rather than references to fixed, unchanging essences. Indeed, the categorisation of American music in racial terms was a deliberate creation of the music industry, when in 1923, acting under the influence of Jim Crow segregation, it decided to market music by blacks and whites separately as 'race' and 'hillbilly' music respectively. Richard Peterson argues that this decision 'had the effect of obscuring the common heritage shared by American-born working-class African Americans and whites and facilitated its separation into distinct streams' (1997: 235). The historical fact that both races contributed to, and drew from, a common rural, working-class heritage became obscured by the categorisation of music along racial lines. This chapter explores the way racial and class factors have informed the blues and country music, and have had particular ethical and aesthetic consequences for their respective representations of the natural world.

## 5.1 The blues and environmental ethics

The blues emerged from the rural experiences of African-American people at the turn of the nineteenth century, and its 12-bar form was formalised by W. C. Handy in 1912 (Oakley 1983: 40-1). Yet, as Giles Oakley observes, a 'mythological or poetic truth' locates the music's origins in the earlier era of slavery. He quotes Booker White, that the 'foundation of the blues is walking behind a mule way back in slavery time' (10). Musically, the blues developed out of the field hollers, work songs and spirituals sung by African-American slaves, through which they articulated their interactions with the natural and built environment within an oppressive social and political system. As African-Americans migrated in large numbers from the rural South to cities such as Chicago, Detroit and New York around World War One, the subject matter of the new city blues shifted away from concerns with life on the farm, to focus on preoccupations with sexual relationships, work, money and leisure.

After World War Two, the young, black, urban audience preferred the newer, up-tempo styles of rhythm and blues, or the sweet blues of B.B. King, and came to see the older country blues styles as reflecting a past of poverty and racial discrimination they hoped they had left behind (Jones 1963: 172). At the same time, the predominantly white, middle-class audience for folk music began to discover the older generation of country blues singers. For this audience, country blues represented rural authenticity beyond the overt commercialism of the popular music industry. To cater to both audiences, John Lee Hooker would play acoustic sets and dress down in dungarees for his white audience, and an electric set dressed in a sharp suit for his black fans (Murray 1989: 132-5). Interest from white audiences led in the early 1960s to country blues singers such as Mississippi John Hurt and the Rev. Gary Davies making records for the first time and performing at folk festivals alongside white folk singers.

The white, middle-class American audience took the blues to be organic, earthy, primal and animalistic because of its agrarian and proletarian origins (Pattison 1987: 37). As Robert Cantwell writes, 'By a weird historical irony, what came springing out of slavery was the most manifestly unencumbered personality on the social landscape, endowed with performance forms that melted the rigid deportments of Euro-American hierarchies in the warmth of an interactive, somatic, aural-oral culture' (1996: 69). According to the Romantic myth of the Noble Savage that underpins such white attitudes to black culture, human beings in their natural state have a healthy, liberated libido, until white society, repressed and repressive, imposes the values of civilisation on it.

Romantic nostalgia for a return to a mythical golden age, before the repression of instinctual life under modernity, informed the 'mouldy fig mentality' adopted by certain white listeners towards the country blues, whereby, as Charles Keil explains, they looked for particular criteria of authenticity in a 'real blues singer': old age, obscurity, correct tutelage and, most important for our purposes, an agrarian milieu. A bluesman, writes Keil, 'should have lived the bulk of his life as a sharecropper, coaxing mules and picking cotton, uncontaminated by city influences' (1966: 35). But as musicologist Allan F. Moore comments, when white audiences take the improvised nature of the blues as a sign of the music's immediacy and naturalness, they ignore the rule-bound, learned conventions from which those improvisations have emerged (2001: 65). The blues, then, is no more 'natural' than any other type of music. Rather, it is equally learned and mediated through technology; as Evan Eisenberg notes, the invention of the ten-inch, 78 r.p.m. disc standardised the blues as a song form in four stanzas (1987: 116). Indeed, twentieth-century technological developments were vital to the development of this form of African-American cultural expression. 'With sound recording', writes Susan McClary, 'a previously silenced group, which had been represented to the broader public (when at all) only through European notation, descriptions, and imitations, could begin to explore and literally to broadcast their own various approaches to self-representation' (2000: 38).

Moreover, African-American writers and scholars have celebrated the blues as an expression of black people's embrace of modernity. The rhythms of the music itself have been widely compared, by musicians and critics alike, to an iconic sound of modern industrial society: the railroad train. Houston Baker writes that 'the dominant blues syntagm in America is an instrumental imitation of *train-wheels-over-track-juncture*' (1984: 8). The blues drew their optimism from this association with the railroads, and with 'technological innovativeness' in general (11). 'Even as they speak of paralysing absence and ineradicable desire', Baker wrote, 'their instrumental rhythms suggest change, movement, action, continuance unlimited and unending possibility' (8). As we will see later in this chapter, Cecelia Tichi makes a similar point about the iconography of the road and the automobile in country music.

The blues have also been viewed by musicians and critics as an expression of black collective resistance to white oppression. In *Development Arrested* (1998), geographer Clyde Woods argues that the blues have provided for the 'working class African American communities in the rural South and their diaspora' a 'sense of collective self and a tectonic footing from which to oppose and dismantle the American intellectual, cultural, and socioeconomic traditions constructed from the raw material of African American exploitation and denigration' (1998: 29). In a later essay, Woods adds an explicitly environmentalist slant to his argument, asserting that an 'African-American environmental ethic' based on support for 'co-operative rural and urban land reform' may be discerned in both the blues and hip hop (2007: 73). In the years after slavery, he claims, working class African-Americans in the rural South developed 'an approach to equitable sustainable development informed by their own philosophical system' (50). The blues and hip hop explore this 'sustainability ethic' by 'making communities sacred and significant beyond their value as commodities' (72). He cites Robert Johnson's 'Hellhound on my Trail' and Tupac Shakur's 'Changes', both of which suggest that 'the expressed desire to escape from the traps of particular places has made the place, the community, and the intellectual sacred' (72).

The role of hip hop in African-American protest will be discussed more fully in Chapter 9. As for the blues, Woods' ambitious claims for an ethics of sustainability in the music are not effectively backed up by detailed reference to specific songs. Yet a closer analysis of blues songs reveals a concern for individual survival as much as collectivism. However, as Susan McClary observes, the celebration of the collective emerges in the blues from the performance of the music rather than from a literal reading of the lyrics. The 'groove that sustains the blues', she writes, 'serves as a conduit linking the body, words, musicians, listeners, and a realm often experienced as sacred'. There is, she adds, 'no transcendence without the body, no individual redemption without the community' (McClary 2000: 36). As we will see in Chapter 10, Charles Keil has theorised extensively on the ecological implications of 'groove' in music.

Roosevelt Charles was one of the country blues singers 'discovered' in the early 1960s, when he was recorded by folklorist Harry Oster in Angola Penitentiary in Louisiana in November 1960. Accompanied by Otis Webster on guitar, his material included the rural work song 'Cane Chopping' and two talking blues about life as an agricultural labourer. 'The Boll Weevil an' the Bale Weevil' is an anti-pastoral which describes the hardships and uncertainties of farming. The song slowly details everyday activities, from breaking up the earth and planting the cotton seed to chopping the cotton: all of these activities demonstrate that 'There's too many if's about farmin''. The farmer has two adversaries, one natural and the other social. Worse than the boll weevil is the bale weevil, the 'bossman with his pencil', who tells the farmer that he must 'try to raise a little bit more next year': the use of the synecdoche of the pencil encapsulates the imbalance of power between intellectual and manual labour. The song ends with the bossman giving the farmer a few dollars to buy clothes for his wife and child at Christmas, while the farmer gets an 'ole suit' patched up after 'the rats done cut a hole in the rear end'. The final line moves from this bitter humour to a sung wail of grievance: 'Woh, Bossman, that ain't no way to treat me'. The soaring transition from talking to singing marks the emergence of political consciousness: Ernst Bloch's concrete utopian impulse in action.

Harry Oster summed up the subject matter of 'Mule Blues' in his liner notes, as a 'dramatic presentation of central elements in the life of the country Negro farmhand – the feeling of affection for his mule, a craftsman's pride in a job well done, the practice of wandering from job to job, the turning to the boss for a few dollars in advance for a big blast, the boss's urging in paternalistic fashion that the ploughman be judicious and save his money for the winter, the yearning for letting off steam in a juke joint after a long hot day in the fields, and finally return to the basic Louisiana diet of red beans and rice'. The song depicts manual farm labour as a source of masculine pride and social identity: 'Watch me when I get on this land'. It also explores the complex relationship between the farmer and his mules, Dot and Carry, who 'don't stand no jive'. In this hierarchy of working relationships, the mules work for the farmer ('Well, my mules is gaited down / Boy, I walk the field all day long'), and the farmer works for the bossman. From an ecocritical perspective, while there is a strong politics of social justice in Charles' talking blues, the 'sustainability ethic' that Clyde Woods finds in the blues appears largely undeveloped.

The emergence of explicit environmentalist awareness amongst blues musicians was indicated by the compilation *Fish Tree Water Blues* (1999), a benefit CD for Earthjustice Legal Defence Fund's campaign 'to save wild salmon, ancient forests and free-flowing waterways'. The album brings together blues songs loosely on the theme of the natural world, but includes only two songs of explicit ecological concern. J. J. Cale's 'Stone River', an elegy for a polluted river, was written especially for the album. The other song, Loudon Wainwright III's 'Hard Day on the Planet', was recorded in 1986, and is a blackly humorous satire on an irrational world out of human control. Included in the list of social ills, from the falling dollar, new diseases, kidnaps and rapes to terrorism, are ecological crises: 'It's hot in December and cold in July / When it rains it pours out of a poisonous sky'. There is no solution to the problem of corrupt human nature: even if a nuclear explosion lets us 'off the hook', the survivors would start up 'all over again'.

'Hard Day on the Planet' brings its element of explicit environmental and political protest into the blues from the allied tradition of the topical folk song, which we will explore in the next chapter. The electric blues also feature as a musical resource for environmental protest in Neil Young and Crazy Horse's *Greendale* (2004). We will see later in this chapter how Young's concern for the plight of American farmers, and the land on which they work, led him to co-found the Farm Aid organisation in the 1980s. *Greendale* extends these humanitarian concerns into an environmental vision that is both necessarily anthropocentric and biocentric. A song suite and accompanying film, Young's study of corruption and its exploitation by the mass media in a fictional West Coast town ends with an endorsement of environmental activism. Inspired by Julia Butterfly and pursued by the F.B.I., eighteen-year old student Sun Green lights out to Alaska to 'save the planet for another day'. The closing track, 'Be the Rain', is an epic indictment of the despoliation of the American West by corrupt government and corporations. An act of biocentric imagination is the key to better care of the environment: 'Be the river as it rolls along / Toxic waste dumpin' from corporate farms'. Salvation comes through an embrace of wilderness ethics, countercultural rebellion and the traditional American figure of the good outlaw. Hence Young's adoption of African-American electric blues boogie, in the style of Jimmy Reed and John Lee Hooker, as the dominant style for the album. Yet *Greendale* also exhibits a playful self-referentiality and knowingness. On her way to Alaska, Sun Green hears her local band, 'The Imitators', playing in 'John Lee's' bar, as if Young is aware of the dangers of pastiche, if it means merely going through the motions, both artistically and politically. In 'Be the Rain', he shouts through a megaphone ('save Alaska! Let the caribou stay!)'), thereby drawing attention to the didacticism of the song. Ultimately, this knowingness makes a political point: by returning to the electric blues, Young and Crazy Horse offer a noisy alternative to stasis and impotence that is grounded in American traditions of popular resistance. Instead of resignation, there is amateurism (with its etymological roots in the Latin for 'love'), improvisation and a necessary call to action: 'We gotta keep going [...] Earth has many enemies / There's much work to be done.'

That the explicitness about ecological issues displayed by white performers such as Loudon Wainwright and Neil Young is largely absent from the African-American blues is not a criticism of the latter. African-American blues lyrics, such as those sung by Roosevelt Charles, which we examined earlier in this chapter, combine imagistic detail with colloquial language, narrative drama and humour; blues singing is emotional release from suffering and oppression. As such, the blues does not have to place itself within a specific political program to be artistically effective or politically insightful. The same may be said for the insights into the relationship between Americans and the natural world that emerges from the traditions of white 'hillbilly' and 'cowboy' music, to which we will now turn.

## 5.2 Country music and rural authenticity

In the mid-1950s, the country music industry responded to the popularity of rock and roll by reflecting on its history, and creating a new sense of heritage and tradition that was eventually institutionalised by the founding of the Country Music Hall of Fame in 1961. The most important conclusion reached in this period of self-reflection was that authentic country music came from the singer's personal experiences of rural life. Authentic performance was identified as a quality that emerges from a close relationship with the land or the natural world. Yet country music is equally a complex construction of urban America. The rest of this chapter explores these contradictions by examining how country music has represented the American rural environment through the conflicting modes of pastoral and anti-pastoral.

Hank Williams said that 'You have to plough a lot of ground and look at the back side of a mule for a lot of years to sing a country song' (Peterson 1997: 210). Bill McCall, President of Four Star Music, confirmed this view: 'In order to understand country music you have to know which end the milk comes out of, perhaps slop a few hogs, and maybe follow a mule down the row behind a plough' (Peterson 1997:

217). As a result of this closeness to the Earth, country music apparently escaped the corruptions of commerciality. In December 1953, *Billboard* published an article entitled 'Genuine Quality Sets Country Field Apart from Tin Pan Alley Music'. There is 'nothing synthetic about country music', the article claimed. 'Unlike Tin Pan Alley writers who create music solely as a mean of earning a living, country music is not manufactured but rather bred by years of living and feeling a situation and then projected by the writer into song' (Peterson 1997: 42).

Writing in the mid-1980s, in the midst of the New Country revival, Lone Wolf Circles upheld this traditional view of the music's rural authenticity and extended it into a model for deep ecology. Country music, he wrote, is 'country folks reacting to the contradictions of the modern world, suffering the pains of the heart, taking solace and inspiration from the sights and sounds and values of rural America. It is unfenced music for unhobbled spirits' (Lone Wolf Circles 1985b: 28). However, the authentic music was currently in decline. 'Every year', he continued, 'there is less country, and seemingly less Country music'. Although the writer does not elaborate on the exact connection between excessive urban development and the decline of country music, he does describe where the authentic music may still be heard. The Country Music Award shows are not likely to feature such music, nor will the 'so-called' Country radio stations, 'except on oldies night'. To find the real thing, the 'True Country' that is 'subjective, personal, and makes no apologies', the reader will have to listen to 'archaic albums that stick with you like old friends and muddy boots, leaving a trail of tales for those who listen' (28).

Lone Wolf Circles' notion of musical authenticity is based on a denial of the contradictions of country music production. By referring to his old country record albums as 'archaic', the deep ecologist has conveniently deflected attention away from the fact that they are also products of the very industrial system he detests. A similar denial informs his description of the music played in 'the dark recesses of small mountain town bars', where authentic country music lives on. He uses two main musical criteria to make his value judgments about true country music. The real songs, he writes, are 'usually sung by their creators, made of tears and sweat and laughter'. Voices are 'at their best unpolished, shaped natural like a crystal. They are sometimes grit-edged with the honesty of real experience' (28). Authentic country music, then, is usually produced by an autobiographical singer-songwriter from a rural area who sings with an untrained voice.

When Lone Wolf Circles reveals a list of his favourite performers, however, the contradictions in his notion of rural authenticity come to the fore. He admires Bob Nolan of The Sons of the Pioneers, Patsy Montana, Emmylou Harris, Riders in the Sky, Rose Maddox and Mary McCaslin as upholders of the authentic tradition of the music (28). However, not all of these singers can be simply described as 'country folks': Emmylou Harris was born into a middle class family in Birmingham, Alabama. Others invented their 'Western' identity. Patsy Montana was born in Arkansas as Rubye Blevins before she reinvented herself as the 'Cowboy's Sweetheart'. Indeed, this invention of a rural or Western persona was typical. Roy Rogers, who co-founded the Sons of the Pioneers with Bob Nolan, was born Leonard Slye, and moved from Cincinnati, Ohio to Southern California to reinvent himself as a singing cowboy (Peterson 1997: 89-91).

As historians of country music have noted, then, the notions of rural authenticity on which the mythology of country music depends are socially constructed, and indeed have been since the early years of the recording industry in the 1920s. Peterson observes that authenticity in country music is an effect achieved by following a set of coded conventions. As he puts it, for musicians, 'establishing the right to speak involves knowing all the conventions of making the music and the nuances of voice and gesture that make their work sound "country"'. Visual signifiers are also vital. 'The boots, the hat, the outfit, a soft rural Southern accent, as well as the sound and subjects of the songs, all help' (218).

Cecelia Tichi describes the popular appeal of the hillbilly and the singing cowboy, both of which became popular figures in country music in the 1920s. 'Both depend on friends and family rather than on written law and officialdom' she writes. 'Both find little use for formal education in their environment, relying on knowledge that comes from a oneness with nature and its creatures. Finally, both feel at a loss in an urban environment, not only unschooled in its appropriate ways of behaviour, but also unrefined in dress, speech, and manners' (Tichi 1994: 67).

Tichi shows how both of these figures, which represented the country music ideal of the working class, rural primitive as a source of spiritual and cultural renewal, were inventions of an urban-based entertainment industry. Up to the 1920s, she notes, country entertainers had worn tuxedos, in a tradition dating back to the nineteenth century. A photograph taken in 1926 shows Dr. Humphrey Bate's string band, dressed in suits against a middle-class interior of drapery and reading lamp. Two years later, the same band is seen in another photograph wearing work shirts, felt hats and farm pants, against a backdrop of fields and a log cabin, and have been renamed The Possum Hunters (67).

This shift to rural chic was instigated by radio announcer George D. Hay, who founded the Grand Ol' Opry radio show in 1928 for Nashville urbanites. Hay had a talent, described by Peterson, for 'crafting a commercial product in the guise of informal amateurism' (77); many artists 'learned how to disguise first-class performance skills behind a rough-cut naive down-home exterior' (80). Hay invented a new dress code for country music that confirmed its identity with the natural world. Such was the popular will to believe in the myth of rural innocence that in 1930, Glen Rice of radio station KMPC in Beverly Hills convinced his audience that The Beverly Hillbillies (who later became the Sons of the Pioneers) rode to the studio on mules.

Country music continues to uphold this notion of rural authenticity by disavowing its urban origins. Its iconography celebrates pre-industrial handicraft technologies, even as the music itself is a product of modern technologies of sound reproduction. As Tichi notes, country musicians view their instruments, even the metal-bodied Dobro, as products of nature rather than of culture. This process of naturalisation, she writes, removes the musicians 'from the world of technological development and social process, which is to say a world of historical movement' (209). However, she also notes the celebration of the automobile, and the ambivalent iconography of the open road, in country music. The exact role played by the opposition between 'culture' and 'nature' in country music is thus complex and ambiguous. The following sections analyse three main tropes around which country music has based its rural mythology – the farmer, the cowboy and the wilderness – in relation to the modes of the pastoral and anti-pastoral.

## 5.3 The farmer in country music

In many country songs, as in the blues and folk traditions with which they overlap, sympathy for the rural poor is displayed through an anti-pastoral mode that tells of their struggles with a hostile nature. The farmer takes the brunt of natural adversity. Harry Smith's *Anthology of American Folk Music* included The Carolina Tar Heels' 'Got the Farm Land Blues' (1930), in which a thief steals a farmer's chicken, his corn crib and his car tires. The forces of nature then also turn on him, when a violent storm, followed by an invasion of boll weevils, destroys his crop. The song ends with the farmer matter-of-factly resigning himself to the superior forces of hostile nature, both human and non-human, and moving to the city. As Smith summed up the song in his laconic commentary: 'Discouraging acts of God and Man convince farmer of positive benefits in urban life'.

Harlan Howard's 'Busted' is a similar catalogue of hardships: the cow is dry, the hen won't lay, and the cotton fields are barren. The only hope is to move on. In Johnny Cash's version, his sparse guitar and bitter vocal delivery add a tone of resilience that refuses self-pity. Cash said of the work songs he recorded on the album *Blood, Sweat And Tears* (1962), which included 'Busted', that they 'brought out voices that weren't commonly heard at the time – voices that were ignored or even suppressed in the entertainment media, not to mention the political and educational establishments'. He was, he said, 'trying to get at the reality behind some of our country's history' (Doggett 2000: 289). Cash's dirty realism necessarily debunked the pastoral myth of American rural life.

Charley Pride, who also recorded a version of 'Busted', was one of the few African-American singers to achieve stardom in country music. If his success suggests the tokenism of the Nashville music industry, it nevertheless shows that the notion of rural authenticity that underpinned the music was not exclusively white. Indeed, to identify country music automatically with whiteness is simplistic. As noted at the start of this chapter, early audiences for traditional country music and blues were inter-racial; while in 1993, the market research company Simmons Research Bureau published a survey which stated that between six and eight million African-Americans, or 24 percent of African-American adults who listen to the radio, listened to country music radio (Centre For Black Music Research nd).

One of eleven children born to poor tenant farmers in rural Mississippi, Pride had all the experiences deemed necessary for acceptance as a country singer (Carlin 1995: 366). In 'I Wonder Could I Live There Anymore' (1970), he sang of a childhood in a 'little three-room house' in the country, of being awakened by a rooster and smelling cornbread cooking on a woodstove. But he also remembers his mother being too busy cooking to talk, and his father working overtime in the fields to pay the grocery bill. The jauntiness of the chorus, with its Chet Atkins' produced Nashville Sound of female backing vocals and uplifting pedal steel guitar, underlines the wisdom of moving to the city: the country may be 'nice to think about it, maybe even visit,' but not a place to live in.

Although these anti-pastoral country songs are ambivalent about the Jeffersonian ideal that virtue is found in the land, they nevertheless tend to endorse his notion that private ownership will make the farmer and his family self-sufficient. By proudly celebrating the independence and survivalist resilience of the family farmer, even these anti-pastoral country songs thus ultimately uphold the long tradition that identifies agrarian life with virtue, or, as J. Hector St. John de Crèvecouer's Farmer John put it in the 1760s, that the 'simple cultivation of the earth purifies' (Tichi 1994: 23).

Other country songs, in contrast, celebrate rural life as a pastoral ideal, and prefer sentiment and reassurance to the realism of Johnny Cash and Harlan Howard. In its imagistic conciseness, Rodney Crowell's 'Long Hard Road (The Sharecropper's Dream)', a number one hit for The Nitty Gritty Dirt Band in 1984, restates the Jeffersonian ideals of hard work and Christian piety. As the narrator reminisces about his childhood in a rundown cabin in the woods, he would rather see 'lightning bugs go dancing in the rain' than all the city lights he has known since.

Many country songs similarly eulogise the natural world as a nurturing home: 'Home, Sweet Home', 'Old Folks at Home', 'Home on the Range', 'My Old Kentucky Home', 'Take Me Home, Country Roads', 'Green, Green Grass of Home' and 'My Tennessee Mountain Home', amongst many others. 'This legacy of American thought', Tichi observes, 'appears in country music's insistence that home is more than a locale and site of family and friends – is actually the essence of the natural world' (1994: 25). In particular, this pastoral side of country music perpetuates the notion of rural Appalachia as what Peterson calls a 'pristine remnant of a bygone natural environment peopled by British American stock, unspoiled by the modernist thrust of urbanisation and industrialisation' (1997: 215).

Yet, as Tichi notes, this myth of home is selective, and is able to maintain itself 'only when problems of race and class do not enter in and disrupt the carefully devised and managed ideal' (1994: 23). By identifying home with nature, country music is able to deny history and to naturalise its conservative values. In his essay 'Farming', Ralph Waldo Emerson argued that moral authenticity lay in rural life, and held up the white, Anglo-Saxon farmer and his family as embodiments of what he took to be the inherent moral goodness of the natural world (Emerson 1894). The strain of Anglo-Saxon suprematism in Emerson's agrarian ideal found its way into the country music produced in the segregated South in the twentieth century. Indeed, Nick Tosches has exposed the overt racism in some country songs (Tosches 1985: 162-217). More typically, however, country's adherence to the pastoral mode allows for racial conflict simply to be ignored.

The corollary of country music's pastoralism is its anti-urbanism. The music, Tichi notes, often explores a negative view of the city, thereby placing 'home's enveloping love and kindness against materialism, social status, hurdles of hierarchy, and all sorts of false value systems', all of which are focused on the city (1994: 35). As Buck Owens sang in 1970: 'I Wouldn't Live in New York City (If They Gave Me the Whole Dang Town)'. In Harlan Howard's 'Streets of Baltimore', covered by Gram Parsons, when a woman's material ambitions drive her to leave the safety of the country, she falls into prostitution in the city. A common country music narrative, Tichi observes, involves a person who 'leaves home in the country, discovers loss, feels homesick, resolves to return' (40). Such anti-urbanism is an American tradition that goes back to Jefferson's fears that a newly urbanised republic would replicate the vice and corruption of European cities.

However, country music also expresses frustrations with the narrowness and claustrophobia of life in rural, small-town America. The music thus presents, in Tichi's words, a 'fundamental American conundrum. To stay at home is to risk feeling dejected and thwarted. But to appreciate home, one must leave it, never to return' (48). In Gram Parsons' 'Hickory Wind' (1974), the city leads him astray, and he yearns for the lost innocence of the country, singing of the desperation of someone caught between two worlds. 'Hickory Wind', writes Tichi, is 'a meditation, not on home, but on the yearning for it'. In life, trouble and lonesomeness are real, and home is an impossible ideal (50). Even the pastoral tradition in country music, then, is ambivalent: it is a complex pastoral (in Leo Marx's terms) that mourns the loss of an ideal. As we have seen, the anti-pastoral tendency in the music, represented in songs such as 'Busted' and 'I Wonder Could I Live There Anymore', refuses even this nostalgia for an idealised rural life.

Historian John Opie points out that the Jeffersonian ideal of the small-scale, family farmer 'has not been the representative American since the turn of the century' (1994: 191). In the 1930s, the rise of agribusiness shifted food production away from smallholdings. Nevertheless, the policy of successive governments until the 1980s was to attempt to rescue the family farmer, as Roosevelt's government tried to do during the Great Depression. In the middle of the 1980s, however, the government began to withdraw support for the small landed farmer to the benefit of large-scale, profit-making operations. Opie writes that the 'inefficiencies of independent farmers, their lack of staying power in hard times, and the decline of their political influence make them the first object of sacrifice' (184). In 1983, the American agricultural economy collapsed when high interest rates and low prices depressed farming in the Mid-West and central South. Banks foreclosed on their loans, throwing many farmers into poverty.

Country songs have tended to condense their response to such complex economic realities into a populist attack on the banks. It was God who made Frankie Miller's 'Black Land Farmer' (1961) a simple man, not a 'banker or legal charmer'. The farmer accepts his lot with resilience, good humour and simple piety; his mind is 'at ease' and he can 'do no harm'. After the farming collapse of the mid-1980s, a flurry of country protest songs also targeted the banks for blame. In Nanci Griffith's 'Trouble in the Fields', on 1987's *Lone Star State of Mind*, the bankers 'swarm like locusts out there turning away our yield', and the family may be forced to sell their new John Deere tractor. But just as 'our parents' survived the Great Depression, so the narrator finds hope in hard work and love for 'our native soil': the rising melody and Griffith's sweet voice is a salve to the hardships of the present.

The Charlie Daniels Band's 'American Farmer' (1985), in contrast, angrily refused a therapeutic response to rural poverty. Dressed on the album cover in Stetson hats and long coats, the band's brash country rock sound, led by twangy guitar and heavy drums, delivers a belligerent condemnation of government policies towards the farmer, specifically of 'selling his land to the big corporations'. In doing so, the song not only addressed the farmer's sense of abandonment, but also, unusually for country music, indicted agribusiness as its cause. If the farmer doesn't work, the chorus reminds the listener, 'then the people don't eat'.

Another politicised country song written in reaction to the farming failures of the mid-1980s was John Denver's 'Let Us Begin (What Are We Making Weapons For)', on the album *One World* (1986). The song was an unusually angry one for Denver, and took the politically ambitious step of linking the farming failures in the United States to the escalation of military spending on the Cold War. After 'four generations of love' on his farm, a farmer has been forced to give up his land. The cause of the problem is the war machine, which is 'taking food out of the mouths of babies'.

At the American Live Aid concert in July 1985, Bob Dylan used his appearance in front of a global television audience to draw attention to the plight of American farmers: 'I'd just like to say that I hope that some of the money that's raised for the people in Africa, maybe they could take just a little bit of it – maybe one or two million, maybe – and use it, say, to pay the, er, pay the mortgages on some of the farms, that the farmers owe to the banks' (Denselow 1989: 246). As Peter Doggett comments, Dylan's timing was bad, given that Live Aid was intended to be about asking Americans to think of others generally less well off than themselves. His speech missed 'the distinction between someone struggling and someone starving' (2000: 581). Nevertheless, it inspired Willie Nelson to launch Farm Aid, an annual telethon in support of the American family farmer (Farmaid.org). In the mid-sixties, Nelson had changed his clean-cut Nashville image by growing his beard longer, putting on hillbilly clothes, changing his repertoire and reinventing himself as, in his own words, 'Your old cotton pickin', snuff dippin', tobaccer chewin', stump jumpin', gravy soppin', coffee pot dodgin', dumplin' eatin', frog giggin' hillbilly from Hill County' (Doggett 2000: 303). Despite this rural fakery, however, Farm Aid shows the emotional and practical commitment of musicians such as Nelson, John Mellencamp and Neil Young to the Jeffersonian agrarian ideal. As Tichi observes, it is significant that the organisation supports 'not international high-tech corporate agribusiness, but farms legitimately of the natural order' (1994: 197). Duncan Webster writes that Farm Aid demonstrates 'the populist power of countering economic conservatism from culturally conservative positions'. He adds: 'In this way reclaiming traditions can be seen as radical – literally, a return to the roots – setting history against capitalism's attempts to deny continuity and against pop's timeless present of consumption' (1988: 169).

Farm Aid's support for organic and local food production is echoed by country songwriter Adrienne Young, who dedicated her album *Plow to the End of the Row* (2004) to, amongst others, Jefferson, Emerson, Thoreau and Wendell Berry. The album sleeve also advertises the Food Routes Network, which aims to promote 'a healthy, responsible lifestyle by supporting locally grown food'. The compact disc was sold with a free packet of wildflower seeds. Despite this involvement in the contemporary politics of food production and ecological restoration, however, the way Young represents farming as a model of virtue in the title track is reassuringly archaic: 'If the mule won't pull then the plough won't go'. Once again, country music disavows the modern technology on which farming depends. The cover art work, depicting old, sepia newspapers, photographs and farmers' almanacs, is also nostalgic for home production, despite the compact disc itself being a product of mass industrial production, and the presence on its sleeve of that key indicator of global contemporaneity, the band's web-site address.

Gillian Welch's 'Annabelle' (1996) continued the harsh tradition of the anti-pastoral in country music, but again went back to the past. The family lease 'twenty acres and one Ginny mule / From the Alabama trust', but gives away half of their cotton and a third of their corn in return for 'a handful of dust'. When her daughter Annabelle dies, the desperation in the narrator's appeal to Jesus recalls the Carter Family. The simplicity of the accompaniment, for two guitars and close harmony singing, creates a sense of space as sparse and desiccated as the landscape evoked by the lyrics. Like Dylan's 'Ballad of Hollis Brown', which will be discussed in the next chapter, Welch's song both indicts the social conditions that create rural poverty, in this case tenant farming, and at the same time surrenders to the inevitability of fate. Acceptance and religious yearning are all the bereaved mother has left.

None of the country songs discussed in this chapter recognises the environmental costs of food production, an issue raised by Adrienne Young's promotion of organic farming methods on the sleeve of her album, but not by the record itself. Opie describes the environmental consequences of government policies which have encouraged the consumption of land rather than its conservation. Farmers have no incentive to invest in long-term conservation because they are dependent on the credit and price system, and so need high yields from their land to stay solvent. 'Less-productive field rotation and summer fallowing', he notes, 'are forgotten under today's conditions of low prices and high interest payments. In addition, tenant farmers, "suitcase" farmers who live in nearby towns, speculators, and other absentees treat the land as a short-term profit base' (192). 'Another disincentive for conservation by local resident farmers', he writes, 'comes from their capacity to compensate for the decline in soil quality by investing in new machinery, higher-yield crops and an increased use of fertilisers and pesticides' (188).

Country songwriters have ignored these issues, as they are clearly damaging to the sentimental image of the benign family farmer in intimate and loving contact with the land, which the pastoral tradition in the music is keen to perpetuate. The anti-pastoral tradition, in contrast, has focused on the human costs of farming failure, while upholding the notion of the dignity of rural labour. Country music's widespread refusal to deal with the realities of farming as a contemporary industry makes such denials possible.

## 5.4 The cowboy in country music

Cowboy songs have been an integral part of the myth of the American West. Cowboys, wrote Barrett Wendell in his introduction to John A. Lomax's collection *Cowboy Songs And Other Frontier Ballads*, 'loved roaming; they loved freedom; they were pioneers by instinct; an impulse set their faces from the east, put a tang for roaming in their veins, and sent them ever, ever westward' (1916: xxii). Lomax's collection included 'Get Along Little Doggies', a song written to work the animals, 'Don't Fence Me In', which celebrated the symbolic power of the western landscape, and 'Blood on the Saddle', which depicted the physical dangers of a cowboy's work.

The cowboys who worked the ranges after the Civil War were mainly Irish- and African-Americans, and their songs were a product of this biracial cultural exchange. Lomax collected 'Home on the Range', for example, from an African-American cook who had worked in a chuck wagon on a Texas cattle trail. By the late nineteenth century, however, these groups were being written out of the history of the cowboy in the name of white Anglo-Saxon supremacism. The mythical cowboy was a racially pure, chivalrous knight, and a figure of escape for the white, urbanised male. 'In the cowboy song as in the minstrel show', writes Robert Cantwell, 'the Anglo-American claimed the sexual and political potency of a black-inflected expressive tradition as his own remanded racial trait, the very command of which assured continued Anglo-American hegemony, thus conjoining its inherent racism to its vaunted self-control' (1996: 72-3).

When the country music recording industry began in the 1920s, most cowboy songs were commercially written by professional songwriters for urban audiences. In the 1930s, Hollywood popularised the singing cowboy in acts such as Gene Autry and Roy Rogers. The singing cowboy was a response to the Production Code, in that westerns could include violence because it was displaced from the contemporary city of the gangster genre into the frontier past. He was also a fantasy figure, an excuse for costuming and wishful fantasies. Ranger Doug of the western group Riders in the Sky described his repertoire as 'escapist lyricism' (Tichi 1994: 116). In his movies for Republic, Gene Autry played himself, as Peterson puts it, 'an open-faced, trusting fellow, rather than the typical rugged bad man turned hero by the love of an Eastern beauty in distress that was characteristic of the genre' (1997: 87). With his horse Champion, Autry was himself a champion of the underdog, saving the exploited and victimised from unscrupulous businessmen.

The development of country music in the 1930s depended on the fact that, despite the commercial success of the singing cowboys, its audience preferred hillbilly to cowboy songs. Even Autry's most successful material was non-western themed. However, although cowboy songs were unpopular, the visual image of the cowboy was appealing. The country music industry therefore combined the best of both genres, by inventing the successful formula, in Peterson's words, of 'dressing the singer of heartfelt hillbilly songs in the heroic cowboy outfit' (82). So acts like Ernest Tubb in the 1940s, and Buck Owens's Buckaroos in the following decade, repackaged hillbilly music without the off-putting style. The association of the cowboy with western optimism, individual freedom and expansiveness was thus combined with songs about the vicissitudes of heterosexual love relationships in rural settings. Musically, the western-influenced guitar tended to replace hillbilly fiddles and banjos, while, by the 1940s, cowboy outfits dominated the Grand Ol' Opry. Needless to say, such bright colours, piping, silver and rhinestones were not working cowboy clothes, but an optimistic fantasy. When Nudie Cohn opened his tailor's shop on Sunset Boulevard in North Hollywood in 1947, it became *de rigueur* for aspiring country artists, from Hank Williams to Gram Parsons, Garth Brooks, Clint Black and Alan Jackson, the 'Hat Acts' of the 1990s.

The cowboy embodied masculine ideals of self-reliance, self-sufficiency, freedom and individualism. He became, in Tichi's words, 'the masculine ideal of personal competence and skill embodied in a single figure who is dependent upon no one else, neither family nor team members nor peer group' (1994: 110). However, although this freedom is mostly a masculine privilege, women in country music often seek self-reliance and self-empowerment as well as the men. The western image, Tichi writes, 'not only frees the female from the gingham and ruffles that signalled her domestic subservience but offers terms approximating the man's' (118-9). Examples of the liberated western woman are Emmylou Harris' version of Tex Owens' 'Cattle Call' (1992), a hit originally for Eddy Arnold, and Tish Hinojosa's 'In the Real West' (2003).

Revisionist western historians such as William Cronon and Patricia Limerick have shown how the individualism associated with the popular myths of the American West aided the national program of economic expansion under the alibi of Manifest Destiny. For these historians, the real values of the American west were greed, ethnocentrism and imperialism, which contributed to a scandalous history written out of the official versions. Cronon and Limerick also argue that traditional western values have proved ecologically disastrous: the ideology of self-reliance is anti-ecological, in that it denies the notion of interconnectedness on which the concept of ecology is based, while Manifest Destiny gave permission for the endless exploitation both of the land and its indigenous peoples (Cronon 1996; Limerick 1987).

The revised image of the cowboy, as Tichi notes, therefore sees him as an agent of corporate growth, 'a corporate hireling, powerless in his lack of property and wealth, a kind of emasculated figure, a two-legged steer in a ten-gallon hat' (115-6). This figure tends to be absent from country music, however. Nevertheless, as Tichi also argues, country and western music is more than simply an 'engine of ideology'. Instead, it promotes the more positive aspects of the western myth, many of which have come to form the basis for environmentalist concerns. To criticise cowboy songs for idealising the outdoor life of the cowboy, she observes, misses the genuine love of the land that they reflect. Moreover, cowboy songs advocate, in Tichi's words, the 'rejection of mere materialism (even in custom boots of exotic hides)', and a sense of the cowboy's dependence on others, including animals (1994: 112). Hence the defence of the cowboy ideal in songs such as Neil Young's 'Are There Any More Real Cowboys', performed as a duet with Willie Nelson on the former's album *Old Ways* (1983), a few years before the two men got together to start Farm Aid. The 'real' cowboy is not the cocaine-sniffer or the one dressed in 'diamond sequins' on television, but the 'working cowboy' who works hard on the land and prays piously for rain. This sense that modern urban civilisation is inferior to the simpler life in the country underpins the environmental activist concerns of musicians such as Neil Young and Willie Nelson that we have encountered in this chapter. Yet their embrace of such notions is not uncritical and naive: 'Old ways', sings Young on the title track, 'can be your ball and chain.'

For John Denver, one of the most commercially successful country singers, the landscape of the American West was crucial to the development of an environmentalist sensibility. It is to Denver's conservationist advocacy we will now turn.

## 5.5 Country music and wilderness conservation: John Denver

Wilderness is less often celebrated in country music than the 'middle', agrarian landscape. The wild, as Tichi notes, is more often 'a state of mind'; the wild rose, for example, symbolising unbounded personal freedom, the unfettered, vitality and beauty of nature (1994: 136). An exception is the work of John Denver, who, until his death in a plane crash in October 1997, was the American popular musician most associated with environmentalist causes, for which he won many awards. He was an advisor to the Wildlife Conservation Society, and a friend of David Brower, Jacques Cousteau and Buckminster Fuller. Christine Smith's book *A Mountain in the Wind* compares Denver to Buddha, Gandhi, Jesus and Mother Teresa, indicating the extent of his charisma (2001:13).

Denver's early success as a country singer was based on nostalgic celebrations of pastoral life in rural America. 'Take Me Home, Country Roads' was a gold record in 1971. *Back Home Again* (1974), his first success in the country charts, included 'Thank God I'm A Country Boy'. After this album, however, Denver refused RCA's request to target a country audience; his decision to adopt a crossover country-pop style instead made him the biggest selling American singer in any genre in the 1970s. It was at this time that Denver began to celebrate the American West in song. Born John Henry Deutschendorf, he had renamed himself at the start of his musical career in the 1960s in tribute to that region. For his record covers, he was often photographed against pristine natural backdrops, evoking the mythic innocence of the nature boy, above commerce and the materialistic temptations of the city. Though Denver's music is occasionally cloying and twee, with safely predictable harmonies and sweetened production, at its best, in 'Rocky Mountain High' (1972), 'Rocky Mountain Suite', on *Farewell Andromeda* (1972), and 'Wild Montana Skies' (his 1983 duet with Emmylou Harris), when his pure voice soars upwards over his ringing acoustic guitar, the music evokes the sense of exhilaration and mythic potentiality of the American western landscape as a 'domesticated sublime' (Cronon 1996: 75).

Denver's love of the American west led him to a wide range of environmental causes in addition to wilderness conservationism. He was a campaigner against nuclear technology, and founded the Windstar Foundation in the 1970s to research renewable energy sources. The proceeds from 'Calypso' (1975) went to Jacques Cousteau's marine research. The song, released on the *Windstar* album, celebrates Cousteau's boat, and the people who work on it, as bringers of enlightenment, working 'in the service of life and the living'. The lesson of the dolphins who accompany the boat is to be 'joyful and loving in letting it be'. The song's yodelling chorus and simple, rousing melody typify the New Age optimism that informed Denver's conservationism.

Denver's range of interests was characteristically eclectic, including the *Tao Te Ching* (the source of the notion of 'letting it be' mentioned in 'Calypso'), the *I Ching*, Zen Buddhism, crystals, astrology, pyramids, yoga, meditation, reiki and Rolfing. He was also involved in self-help therapeutics, including Erhard Seminar Training (EST), which claims to eradicate false identities to reveal the real self beneath. Like many New Agers, Denver was also interested in pioneering technologies. Unusually, in his case, this extended to both aviation and the space program. He wanted to be the first civilian in outer space, and won a NASA Medal for Public Service (Smith 2001: 99, 51, 107).

In making such enthusiasms part of his environmentalist vision, Denver resembled John Cage, who shared his faith in the cornucopian potential of post-scarcity economics. Indeed, the link between the two men, whose aesthetics were so far apart, was R. Buckminster Fuller. Like Cage, Denver knew Fuller, and advocated his belief that social change comes about through small, individual actions. Fuller called himself 'Trimtab', after the miniature rudder on a ship, which has a disproportionate effect compared to its size. Denver dedicated the piano ballad 'What One Man Can Do', on 1982's *Seasons of the Heart*, to Fuller, a 'friend to all the universe' and 'grandfather of the future'.

Like Fuller, Denver also rejected party politics for grassroots activism, community projects and lifestyle politics. Social change, he believed, would come about through individuals acting as global citizens, rather than through politicians or non-governmental organisations. 'People on an individual basis will make changes, not protestors or lobbyists', he told a reporter in the early 1970s. 'People who do what they really know to be right or true. Little things. In traffic, in grocery stores, you let somebody else in front of you. That's peace' (Smith 2001: 41).

In 'World Game' (1983), Denver summarised his ideas about politics and public service in an upbeat pop-reggae number. A future based on 'using less, doing so much more' can be achieved through co-operation rather than competition: instead of 'you against your neighbour', it should be 'I and I'. Like the musicians discussed in Chapter 10, Denver adopts world music stylings as a signifier of global unity and ecological consciousness. In an article in the *Wind Star Journal* in 1986, he wrote: 'We can no longer exist in the contradiction between what we say in our words and deeds as an expression of our fear and separation, and what we hear in the cry of our hearts expressing connectedness and wholeness and the recognition that we are One' (Smith 2001: 18).

Unfailingly optimistic and positive, Denver's music is nurturing rather than challenging. More so than the protest folk tradition within which he began in the 1960s, and which is the subject of the next chapter, the country music Denver adopted allows for expressions of sentiment. As Tichi observes, country music enabled the rugged individualist American male, conditioned to be distrustful of language and emotional expression, to be sentimental in song. 'Through song', she writes, 'the individualist can break his silence and release a flood of thought and feeling'. In this way, the 'Wild, Wild West in country music serves to free the self-reliant American from an imprisoning solitude of the American heart' (1994: 130). The sentimental pastoral tradition in country music, exemplified by John Denver, which sings of the natural world as a home, is an important artistic expression of what Ernst Bloch called 'abstract' utopian yearnings. However, to the extent that Denver adds a political dimension to such desires, his songs may fit more closely into Bloch's more progressive notion of 'concrete' utopian expression. However they are conceptualised, such pastoral trends in American country music run counter to the alternative tradition of anti-pastoral in the music, identified in this chapter with Johnny Cash, which focuses more unsentimentally on the harsh realities of human relationships with the natural world.

# Chapter Six

# Folk

When folksingers Pete Seeger and Malvina Reynolds turned to environmentalist advocacy in the mid-sixties, their new political direction was a logical development of their earlier involvement in the Civil Rights movement, and, before that, the Popular Front during World War Two. Seeger's *God Bless the Grass*, released in January 1966, was the first album in history wholly dedicated to songs about environmental issues, many of which were written by, or co-written with, Reynolds. She had earlier written 'Little Boxes', the satire on middle-class, suburban conformity that was a hit for Seeger in 1963, and which stands as a transition to the more explicitly environmentalist songs on *God Bless the Grass*. This chapter will analyse this album in detail, and then consider developments in the history of the topical folk song in the succeeding decades as they relate to environmental issues. It ends with an examination of an alternative tradition in American folk and blues music which in the 1990s came to be known as the 'American Primitive'. Viewed ecocritically, this music is less concerned with explicit environmental protest than the tradition headed by Pete Seeger, but instead views the natural world in Romantic terms as a place of enchantment and spiritual revelation.

## 6.1 The rise of environmental folk music: Pete Seeger

Seeger's *God Bless the Grass* both continued, and deviated from, the tradition of topical folk song in the United States established during the period of the Popular Front against fascism. The Popular Front was created in 1935, when the Seventh World Congress of the Communist Party called for the workers and the bourgeoisie to unite across class lines against the greater, common enemy of fascism. Aesthetically, the Popular Front advocated the aesthetic values of socialist realism which had influenced left-wing folk music in the United States since it was adopted as the official policy of the Communist International in 1928. Socialist realism, or 'proletarian realism' as it was called in the United States, had the direct social purpose of supporting the working class in its struggle against the bourgeoisie. Lenin had decreed that art must 'penetrate with its deepest roots into the very midst of the labouring masses. It must be *intelligible* to these masses and be loved by them. It must unite the feeling, thought, and will of these masses; it must elevate them' (italics in the original) (Denisoff 1971: 15). For Lenin, art should be a weapon: classical music only soothed the masses, he said, whereas art should 'hit them on the head' (15). Under Lenin's influence, many members of the American Communist Party chose folk music as the people's art form. Party member Mike Gold wrote that art should be for 'lumberjacks, hoboes, clerks, sectionhands, machinists, harvesthands, waiters – the people should count more than the paid scribblers' (15). Folk 'songs of persuasion' were thus works of agitation-propaganda, in which artistic form was considered secondary to content, and language was intended to be a transparent means of communication. The musicologist Charles Seeger, father of folksingers Pete, Mike and Peggy, played a key role in promoting proletarian realism in classical music in the 1930s. His son Pete did the same for folk (Davis 2002; Reuss 2000).

As already mentioned, when Pete Seeger and Malvina Reynolds became involved in environmental advocacy in the early sixties, their politics were informed by the same ideologies of nationalism, populism and criticism of big business as their earlier Popular Front commitments. The song collection which Seeger put together with Woody Guthrie in 1940, *Hard Hitting Songs For Hard Hit People*, contained songs such as 'Mister Farmer', written in support of a strike by dairy farm workers over the low prices given by the corporate owners for their milk (Greenway 1960: 214). Advocacy of conservationism, however, potentially conflicted with this celebration of the industrial worker which was central to the proletarian realist folk tradition. This celebration had a long history in American folk music, from the Industrial Workers of the World to Woody Guthrie. In the early decades of the twentieth century, the Wobblies popularised political folk song particularly through union organiser Joe Hill. These proletarian realist songs celebrated male struggles in heavy industries such as mining. Aunt Molly Jackson, a coal miner's wife from Appalachia, sang 'Which Side Are You On', written during a coal strike in Harlan County, Kentucky in 1932, to protest conditions in the mining industry and call for the unionisation of mine workers (Greenway 1960: 169). As Georgina Boyes observes of the similar repertoire of British folksingers Ewan MacColl and A. L. Lloyd in the 1950s, the industries worth researching and writing about 'were never catering or nursing, hairdressing or office work, and only the heroic was celebrated' (1993: 240).

At the centre of Woody Guthrie's songs in the 1930s, however, was a rural nostalgia that partly distanced them from these materialist tendencies in the Left. As Robert Cantwell writes: 'Whereas the leftist program was essentially secular and historical, grounded in economic theory and tending toward more or less radical reform, Guthrie's vision – or, rather, the vision around which he and the other cultural seekers and idealists converged – was essentially pastoral and mythical, echoing Christian eschatology and rooted in the Gospel according to Matthew' (1996: 137). Nevertheless, Guthrie believed that his pastoral vision of 'Pastures of Plenty' would be brought about by industrialisation and modernisation. America was a democratic land of material plenty prevented only by an artificial scarcity which industrial development would remove. Guthrie's songs thus celebrated the big public works of President Roosevelt's New Deal, such as the building of the Grand Coulee Dam on the Columbia River. In the early 1940s, Guthrie was hired by the Bonneville Power Administration to write songs celebrating these new construction projects (Carriker 2001).

The Folk Revival of the late 1950s continued this workerist tradition. Phil Ochs' 'Automation Song' returned to John Henry's mythic protest against the replacement of human labour by the machine. The male worker, left only with 'the muscles on my arm and the sweat upon my back', is walking down the road, jobless and with nowhere to go. Gordon Friesen, in the liner notes to Ochs' album *All the News That's Fit to Sing* (1964), commented that the 'same feeling expressed here has led Phil to write songs for, and several times visit, the impoverished mining families in eastern Kentucky made jobless by automation'.

Both Pete Seeger and Malvina Reynolds inherited this proletarian realist celebration of the heroic industrial worker. Seeger's first group, the Almanac Singers, formed in December 1940, did benefits for strikers at which they sang songs such as 'Talking Union', about the struggles for unionisation of industrial workers such as miners and automobile workers. Reynolds had also been involved in socialist causes. By the early 1960s, however, both singers had managed an uneasy reconciliation of their Old Left notions of authenticity through manual labour with their turn to environmentalist advocacy. Reynolds thus continued to write protest songs in support of the rights of industrial workers at the same time as she voiced her concerns on environmental matters. In 1963, Seeger recorded 'Mrs. Clara Sullivan's Letter', which Reynolds had written in support of miners in Scuddy, Kentucky, who were striking, as the song put it, 'for jobs at decent pay'. The chorus noted that 'there's no better man than a mining man' (Seeger 1993: 110). Here, allegiance to the working class was predicated on a denial of the wider environmental effects of the mining industry. In contrast, the last verse of 'Cement Octopus', on *God Bless the* Grass, which Reynolds wrote for the protests against the proposed construction of a freeway through Golden Gate Park, San Francisco in 1964, took a more environmentalist line, by recognising that constraints on development may cause unemployment, but suggesting a solution in the redeployment of workers into more environmentally sound jobs: 'The men on the highways need those jobs, we know / Let's put them to work planting new trees to grow'. In this way, Reynolds' song successfully negotiated the problem of reconciling the new environmentalism with traditional leftist workerism.

For both Reynolds and Seeger, this growing environmentalist concern for the destructive effects of industrialism came in spite of their membership of the American Communist Party, with its faith in progress though industrial development. Yet their allegiance to Communist Party orthodoxies had always been loose. Reynolds left the Party in the late 1940s, because, as she commented in 1977, its leadership 'had no concept of what I was doing or of what effect it would have' (Lieberman 1989: 78). Although Seeger continued to be a member of the Party, his group the Almanac Singers, as Cantwell writes, were not driven by its artistic agenda but by a 'sheer love of a dreamed-of folk America, or more precisely of American folks, whom few in the group had ever known up close' (Cantwell 1996: 131). Seeger described his next group, The Weavers, which he founded in 1948, as 'political in the broad sense – we weren't Progressive Party, Communist Party, or even Peace Party singers, but we realized that the human race was in a bad situation, and looked on music as part of the on-going struggle to get the human race together, banish war for ever more, bring peace, justice and jobs for all, and all those nice-sounding things' (Denselow 1989: 13).

Speaking in 1963, Seeger again qualified his allegiance to Communism, writing that it derived not from his reading of Marx and Lenin, but from his childhood interest in Ernest Thompson Seton. The Canadian naturalist, he wrote, 'held up the Indian as an ideal, for strength and morality and selflessness, and in tune with all of nature. Anthropologists call this period of human history "tribal communism". I think next time some character asks me, "Are you a communist?" I'll answer, "Oh, about as much as the average American Indian" (Seeger 1963b: 67). Like Woody Guthrie, then, Seeger's political values were informed more by pastoral nostalgia and American nationalism than by the dogmatic economic materialism of the Communist Party. His pseudonym for his regular column in *Sing Out!* magazine was Johnny Appleseed, the legendary patron of American agriculture and horticulture who dispersed apple seeds as he moved westwards across the North American continent. It was this sentimental adherence to a pastoral ideal of America that smoothed Seeger's transition into environmentalism.

The muscular Christianity of Guthrie and Seton was another important factor in Seeger's adoption of environmental causes. 'As a kid', he wrote in 1993, 'I'd been a nature nut. Age 15 and 16, I put all that behind me, figuring the main job to do was to help the meek inherit the earth, assuming that when they did the foolishness of the private profit system would be put to an end. But in the early '60s, I realised that the world was being turned into a poisonous garbage dump. By the time the meek inherited it, it might not be worth inheriting. I became an econik' (Seeger 1993: 201). For Seeger, then, environmentalism, by mitigating the damage inflicted on the world by the capitalist system, was a necessary prerequisite for a future socialist society that would itself be a fulfilment of Christ's humanist, egalitarian vision.

Seeger first made public his environmentalist concerns as early as 1958, when he told a radio interviewer in Cleveland, Ohio: 'Look at the waste we make of our rivers, beautiful clear streams like the Hudson which flows past my door – an open sewer! [...] A river which was once clean and clear – Indians speared fish twenty feet down – is now an open sore. Nobody swims in it; you go on a boating trip, you just don't look down' (Dunaway 1985: 285). In the same year, he wrote 'Oh, Had I a Golden Thread', released on *Strangers and Cousins* in 1965, in which he sang of 'a land of parks / Where people can be at peace'. His concern for pollution could be seen in his wish that the 'the land will be sparkling clean'.

In October 1963, Seeger mentioned his environmentalist views for the first time in his Johnny Appleseed column in *Sing Out!* While chopping firewood, he wrote, he had made up new verses to 'Take This Hammer': 'I don't want no s..t (sic) filled river; past my door, no, past my door. I don't want your litterbug highway, through my land, no, through my land. I don't want no d..n (sic) fool strontium, in my sky, no, in my sky'. These verses, he added, were 'irrelevant, I suppose, to all but me' (Seeger 1963c: 65). This diffident, apologetic tone had disappeared completely by June 1965, when he recorded *God Bless the Grass*.

When the album was released in January 1966, the freedom and topical song movement was declining in popularity, due, culturally, to the rise of rock music, and, politically, to the growing schism within the Left over race. *God Bless the Grass* thus marked a new social and political cause for Seeger at a time when he was beginning to feel marginalised by the leadership of the Civil Rights movement. He and his wife Tochi had taken part in the protest march from Selma to Montgomery in March 1965 on the direct invitation of Dr. Martin Luther King, Jr. However, Seeger sensed the new direction that African-American politics were taking. Late in the same year, the Student Non-Violent Co-ordinating Committee had begun to purge itself of its white, northern, middle-class elements to become an organisation led by African-Americans for African-Americans. Seeger's prominent role in the Civil Rights movement was at an end (Dunaway 1985: 236ff). Biographer David Dunaway notes that Seeger was stung by criticisms that African-American organisers had made of his role as a paternalistic outsider in the southern struggle for Civil Rights. Moreover, he suggests that political despair over the failure of Seeger's integrationist dreams, and his exclusion from the movement with the rise of black separatism, were important motives behind his adoption of environmental causes. Seeger began to spend more time on building the *Clearwater*, the replica of a nineteenth-century sloop which, since its launch in May 1969, has been used by the Clearwater organisation to campaign for the cleaning up the Hudson River (Wilkie 1969: 2-3).

More positively, environmentalism was for Seeger a cause which, as Dunaway puts it, was 'closer to home, one he had understood instinctively as a child' (1985: 280). Indeed, it was the first cause that Seeger had adopted that was located in his own personal experience. Seeger's involvement in environmentalism thus confirmed the shift in leftist activism at this time towards working in one's own community. As the sleeve notes to the 1974 *Clearwater* album observed, 'At some point in the mid-sixties, he decided that it was time to stay home [...] to fight for his ideals in a somewhat smaller sphere [...]' But Seeger's turn to local conservationism in 1965 was not a retreat from global politics, in that it also coincided with his joining the newly formed protest movement against the Vietnam War. Indeed, a year after *God Bless the Grass* he released his famous anti-Vietnam War protest song 'Waist Deep in the Big Muddy'. Crucially, Seeger saw the Vietnam War as the murderous product of the same military-industrial complex that was responsible for polluting the Hudson River near his home. Dunaway records an incident in which a conservative board member asked Seeger not to sing 'Big Muddy' at a *Clearwater*-sponsored event, saying to him, 'We're singing about the water. Can't you stay away from all that Vietnam stuff?' Seeger replied that 'all these subjects are tied together. You know why we don't have money to clean up this river? Guess who takes the big bite out of the tax dollar?' (292). For Seeger, then, the global and the local coincided.

The main environmental concerns of *God Bless the Grass* were with pollution, over-development and resource wastage. This focus on 'quality-of-life' issues was mainly anthropocentric: human happiness and well-being were under threat from these destructive forces. The new environmental protest songs thus reflected the main areas of social concern for those mainly white, middle-class Americans who made up the emerging environmental movement. Environmental hazards were seen as threats to a pastoral America whose imperilled green spaces were natural resources that existed to fulfil human needs for leisure and recreation. However, in a handful of songs, a more biocentric perspective also began to emerge, as we shall see (Eckersley 1992: 57; Pepper 1996: 48-53).

Malvina Reynolds wrote the four consecutive tracks on side one that establish the album's environmentalist theme: '70 Miles' (co-written with Seeger), 'The Faucets are Dripping' (published as early as 1959), 'Cement Octopus' and 'God Bless the Grass' itself. The album embodies the proletarian realist aesthetic of clarity and accessibility in form, and optimism in content. 'Seventy Miles' uses bathos to make its point: 'Seventy miles of open bay – it's a garbage dump'. 'Cement

Octopus' develops its extended metaphor for highway construction with wit and sardonic humour. The refrain is a direct appeal for collective action: 'Oh, stand by me and protect that tree / From the freeway misery'.

The liner notes to the album were written by liberal Supreme Court Justice William O. Douglas, proponent of the Wilderness Act which had been passed in 1964. Douglas' essay asserts the value of wilderness for American national identity, and argues that wilderness areas should not be overdeveloped to serve recreational uses. 'There are dollar values in our mountains to be exploited', he wrote. 'But a tree is measurable not only by its board feet or its cellulose content, but by its beauty, the wildlife it shelters, the biotic community it nourishes, and the watershed protection it gives'. This acknowledgment of the need to conserve wilderness for reasons other than its commercial value suggest an extension of environmentalist concerns beyond an exclusively anthropocentric focus on human welfare. Two songs on the album reflect this direction. Seeger wrote of 'Coyote, My Little Brother', written by Native American folksinger Peter LaFarge: 'Pete told me that the Indian name for the coyote was "Little Brother". His song used the fate of the coyote to symbolise what modern man was doing to all of nature (whereas the Indian lived in harmony with nature).' (Seeger 1967: 54) This quietly understated lament for the declining coyote thus shows a biocentric sensibility growing out of Seeger's sympathy for American Indian cultures.

A similar perspective was also developed in 'The People Are Scratching', a witty ballad written by Seeger, Ernie Marrs and Harold Martin about pollution of the food chain, which shows the influence of Rachel Carson's *Silent Spring*. Its nine verses tell a story of complex ecological interrelationships: a cold winter kills the grass, so rabbits start eating the bark off the trees. The farmers poison the rabbits; the hawks and owls die from eating them, so the field mice multiply and spread to the town; city people poison the mice, and their cats die from eating the poisoned mice, leaving the people scratching from the fleas, which multiply in the absence of their natural predator. The moral of the story is for all small creatures to 'Stay clear of the man with the poisonous hand'. Even though the factor that initiates the environmental collapse is a natural one, it is human stupidity that escalates a problem that could have be solved, not by poison, but simply by feeding 'a few bales of hay' to the starving rabbits.

In the winter of 1965, after recording *God Bless the Grass*, Seeger began to concentrate his energies on building the *Clearwater*. 'My Dirty Stream (The Hudson River Song)' thus highlighted the cause of river conservation that was to be the focus of his political activism for the next decade. The song, first published in 1961, is an elegy for the lost perfections of an Edenic nature. As already noted, Leo Marx defined the 'complex pastoral' as a mode which works 'to qualify, or call into question, or bring irony to bear against the illusion of peace and harmony in a green pasture' (1964: 25). In 'My Dirty Stream', as the narrator sails down the dirty Hudson, the lilting melody and gently swinging rhythm of the banjo, suggesting the 'tacking to and fro' of the boat, evoke a pastoral ideal ruined by what Marx calls the 'counter-force' of industrial pollution. The river is 'crystal clear' at its source in the northern mountains, but downstream is polluted by chewing gum wrappers dropped 'by some hikers to warn of things to come'. Retaining the optimism of the proletarian realist tradition, however, Seeger keeps the dream that some day the river will 'once again run clear'.

Seeger's use of the pastoral is not merely a late Romantic nostalgic yearning for an impossibly utopian relationship with nature. Instead, it is future-orientated and radicalised by his socialist politics. It is, in Bloch's terms, a concrete utopia. 'My Dirty Stream' reconciles its criticism of industrialism with a generalised sympathy for the working-class reminiscent of Seeger's Popular Front politics of the 1940s. The workers at the Consolidated Paper Plant have 'honest hands', he sings. But the song singles out neither the bosses nor middle-class consumers for blame. Instead, it takes a more liberal view, using the third person plural to attribute collective responsibility for, and complicity in, environmental degradation: 'Five million gallons of waste a day / Why should we do it any other way?' In contrast, Malvina Reynolds' songs on *God Bless the Grass* take a more orthodox leftist view of political responsibility for environmental degradation: 'The Faucets are Dripping' blames absentee landlords for the wasted water that leads to dry reservoirs; the 'cement octopus' of Californian highway construction emerges from the state capital Sacramento; in 'Seventy Miles', permission to dump garbage in San Francisco Bay is granted by the 'city fathers'.

As already mentioned, Seeger's involvement in river conservation can be seen as part of a general shift in left-wing activism at this time towards working in one's local community. But while the autobiographical element in 'My Dirty Stream' ('But I live right at Beacon here') is unusual for Seeger, the notion of self that the song constructs is typically outward looking and social. When the song was published in *Sing Out!* in July 1969, it was described as Seeger 'sharing a private dream warmly and intimately' (Seeger 1969: 1). The private dream had become a public, political concern.

In July 1965, one month after he recorded *God Bless the Grass*, Seeger, with fellow organisers Alan Lomax and Theodore Bikel, was outraged when Bob Dylan, dressed in a black leather motorcycle jacket, gave a concert of loud, electric rock music with the Paul Butterfield Blues Band on the closing night of the Newport Folk Festival. Within a year of Dylan's ground-breaking performance, writes Greil Marcus, 'what had been understood as folk music would as a cultural force have all but ceased to exist' (1997: 13).

The new rock music challenged many of the long-held proletarian realist values of the topical folk song tradition, those values embodied in *God Bless the Grass*. Most importantly, in folk circles, the collective response of the audience was considered more important than the artistic demands of the individual performer. Communal singing established a shared identity around a political cause. Dunaway describes the relationship between Seeger and his audience as being like a concave mirror: he 'focused his listeners' admiration back out into the balconies, inspiring people with an image of themselves as better (more tolerant, compassionate, international) than they were. He kindled their hopes' (1985: 228). Seeger himself revealed his aesthetic principles in a letter to a festival organiser in Japan, in which he wrote that 'we need songs with strength to make cowards stop fleeing, turn around, and face the future with a breath of courage'. His songs, he said, 'must be so good that they reach out to 190 million Americans' (270).

Questions of artistic quality were thus paramount in Seeger's mind. In his introduction to Woody Guthrie's *California to the New York Island* (1960), he elaborated on his artistic aims by listing the three qualities that singers could learn from Guthrie's performance style. The first was his 'matter-of-fact, unmelodramatic, understatement throughout'. The second was 'simplicity above all – and getting the words out clearly. They are the most important part of the song'. The third quality was 'irregularity'. He explained: 'to avoid a sing-song effect, from repeating the same simple melody many times, Woody, like all American ballad singers, held out long notes in unexpected places, although his guitar strumming maintained an even tempo. Thus no two verses sounded alike. Extra beats were often added to measures' (Heylin 2000: 82).

Proletarian realism was not lacking in artistry, then. Nevertheless, the form had its aesthetic limitations. Dunaway notes that, in the late sixties, Seeger continued to maintain his optimistic public persona even when privately he became more sombre and pessimistic as the Vietnam War intensified. 'He had endured and persisted', Dunaway writes, 'even when his optimism weakened, for revolutionists have to say they will overcome, even when they know they may not' (1985: 272). It was the repressive limits of this official optimism, however, that Dylan's new repertoire exposed. He had written topical protest songs between January 1962 and June 1963, but then began to rethink his involvement in the movement. Discussing *Another Side of Bob Dylan*, he told journalist Nat Hentoff in June 1964, 'There aren't any finger-pointin' songs in here […] You know – pointing to all the things that are wrong. I don't want to write for people anymore. You know – be a spokesman […] From now on, I want to write from inside me […]' (Heylin 2000: 160).

Dylan's rejection of the protest song in this period may be seen as part of a growing body of dissent within folk music circles about the political efficacy of proletarian realism. In particular, the belief grew amongst some critics and performers that listeners to topical protest songs, rather than having their consciousnesses raised, might only be having their consciences salved. When *Freewheelin' Bob Dylan* was released in May 1963, critic Paul Nelson criticised the aesthetics of the topical protest song movement as practiced by singers like Dylan and Phil Ochs. He commented later: 'We were sort of the anti-topical-song people, not because we disagreed with it politically but just because we thought it was such shitty art, y'know. These songs were like fish in the barrel stuff […] It's like patting yourself on the back music, it just seemed so obvious and not particularly well done' (Heylin 2000: 123).

By 1966, Julius Lester, an African-American songwriter friend of Seeger's, had also come to doubt the political efficacy of the topical folk song movement. The church bombing in Birmingham, Alabama in September 1963, which killed four black children less than a month after the March on Washington, was for him a political turning point. In an article in *Sing Out*! entitled 'The Angry Children of Malcolm X', Lester wrote: 'Those northern protest rallies where Freedom Songs were sung and speeches speeched and applause applauded and afterwards telegrams and letters sent to the President and Congress – they began to look more and more like moral exercises. See, my hands are clean. I do not condone such a foul deed, they said, going back to their magazines and newspapers, feeling purged because they had made their moral witness' (1966: 24-5). For Lester, then, as for Paul Nelson, protest folk music had merely come to promote a complacent listening position.

The 1965 Newport Folk Festival thus saw Dylan's new aesthetics of noise, provocation and interiority clash with the proletarian realism of Pete Seeger and Malvina Reynolds. Seeger did not object to his use of electric guitars *per se*, as Johnny Cash had used electric guitar accompaniment at the previous year's festival. Rather, he thought that the volume was too loud, and it was difficult to hear Dylan's words. As he commented later: 'It wasn't a real sound check. They were tinkering around with it and all they knew was, "Turn the sound up! Turn the sound up!" They wanted to get volume' (Heylin 2000: 211). As a result, when Dylan performed 'Maggie's Farm', Seeger said, 'you couldn't understand a word, because of the distortion' (Seeger 1993: 173). Joe Boyd, who engineered Dylan's sound at Newport, acknowledged that the volume was high, but pointed out that the sound mix was deliberate. By the standards of the day, he said, 'it was the loudest thing anybody had heard. The volume. That was the thing, the volume […] It was powerfully, ballsy-mixed, expertly done rock and roll' (Heylin 2000: 211).

Newport 1965 can thus be seen as a clash between the old Apollonian folk and the new Dionysian rock: the wordy and the ballsy. Seeger's music was genteel, ascetic and emotionally contained; appealing to the superego rather than the id, it was rational message rather than disruptive noise. In contrast, Dylan's new music and performance style signified rebellious individualism, the pursuit of pleasure, and the division of society into the hip and the square, the young and the old. Moreover, in rock music, as Joe Boyd understood, the voice is part of the mix, the music's affect deriving from the overall sound rather than from the foregrounding of lyrical content. In this sense, Dylan's new music was a threat to the entire social and political project of the topical song movement, his performance at Newport in 1965 the symbolic moment in which the Puritanism and populist certainties of the folk movement began to give way to the new, hybrid forms of 1960s rock music and countercultural protest.

The ways in which Dylan was reinventing popular music may be sensed in the positive reaction of Jim Rooney, a member of the Cambridge, Massachusetts folk scene and later a record producer, who reviewed the festival for *Sing Out!* He described how, at the start of the day, Pete Seeger had played a tape of a newborn baby, and asked the audience to sing to it, and 'tell it what kind of a world it would be growing up into'. Rooney commented that Seeger 'already knew what he wanted others to sing. They were going to sing that it was a world of pollution, bombs, hunger, and injustice, but that PEOPLE would OVERCOME'. Rooney compared this attitude to the new direction in which Dylan was taking popular music, and asked: '(But) can there be no songs as violent as the age? Must a folk song be of mountains, valleys, and love between my brother and my sister all over this land? Do we allow for despair only in the blues? … (That's all) very comfortable and safe.' Dylan was, for Rooney, the only performer at the festival who 'questioned our position' and 'shook us' (Nelson 1965: 31-2). He sensed a new aesthetic that was urban rather than rural ('mountains, valleys'), and rejected the sentimental Popular Front insistence on optimism, collective solidarity and predigested dogma (Seeger 'already knew what he wanted others to sing').

Greil Marcus elaborates on the way Dylan's music allowed for an exploration of individual subjectivity prohibited by the folk revival's insistence on collectivism and ideological orthodoxy. Protest songs such as the ones Dylan wrote in his early career were, he writes, 'pageants of righteousness' in which there was no room 'for the selfish, confused, desirous individual who might suspect that his or her own story could fit no particular cause or even purpose' (Marcus 1997: 27).

After what he saw as the personal humiliation of Newport, Seeger resigned from the board of directors of the festival, and temporarily dropped his *Sing Out!* column, before resuming it in December 1966. The new rock culture was rejecting as outmoded and restrictive the Puritanism that underlay Seeger's musical aesthetics. In *God Bless the Grass*, then, Seeger's new found conservationism had been expressed through what for the newly emerging rock audience of the time appeared to be artistic conservatism. Rock music also raised new questions about environmental ethics and aesthetics, as we will see in the next chapter.

The opposition between Seeger and Dylan should not, however, be overstated. By 1967, Dylan had returned to acoustic folk music with *John Wesley Harding*, and subsequently wrote topical protest songs such as 'George Jackson' (1971) and 'Hurricane' (1976), as well as more general political songs such as 'Political World' (1989). Nevertheless, for Seeger, the debacle of the 1965 Newport Folk Festival had the immediate effect of giving him further impetus to dedicate himself to his *Clearwater* project, which was a practical engagement with the environmental concerns raised in *God Bless the Grass*.

## 6.2 Folk music and environmental protest since the 1960s

Like the tradition of topical folk song of which it was a part, the music of environmental protest did not disappear with the rise of rock music in 1965, even if folk music as a whole lost some of the cultural prestige it had enjoyed during the Folk Revival. Folk songs were an integral part of the *Clearwater* project, and Seeger carried on voicing his concerns in songs such as Bill Steele's 'Garbage' (1969) and 'Throw Away That Shad Net' (1975), otherwise known as 'The PCB Song' (Seeger 1993: 211). Don Maclean was also a member of the *Clearwater* crew in the early 1970s. His 'Tapestry' (1970) combined a sense of environmental apocalypse with an ecocentric celebration of the web of life that recalls Seeger and Marrs' 'The People Are Scratching' or LaFarge's 'Coyote, My Brother' on *God Bless The Grass*: 'every thread of creation is held in position / By still other strands of things living'. In contrast to this delicate beauty, the 'smouldering cities' are 'so gray and so vulgar': industrialism and urban living are unnatural impositions on the web of life.

The early 1970s saw a spate of environmental protest songs in both folk and rock styles: a selection of the latter will be discussed in the next chapter. In the folk scene, Tom Paxton wrote his conservationist anthem 'Whose Garden Was This' for the first ecological teach-in at Northwestern University in 1970. The song is a melancholy, minor chord lament, decorated with recorders and delicate strings, for a natural world that has been totally destroyed. All that remains are simulacra: the narrator has seen 'pictures of flowers' and heard 'records of breezes', but longs to experience the real thing.

John Prine's 'Paradise' (1971) explored the sense of place achieved by Seeger in 'My Dirty Stream', by relishing the specificity of American place names. The narrator's childhood haunt of the Green River in Muhlenberg County has been destroyed by coal mining, and the song names names: 'Mister Peabody's coal train has hauled it away'. The use of concrete detail by songwriters like Prine, Seeger and Reynolds can foster a sense of place that is a vital early step in developing an ethics of care for one's environment, or what Lawrence Buell calls 'critically aware place-connectedness' (2001: 66). As Buell also notes, indignation and betrayal are key tropes in narratives of wilderness preservation.

Environmental protest folk song continued into the 1970s in the work of Earth First! supporter Walkin' Jim Stoltz, whose songs celebrate the wild rivers, mountain trails and wild animals he encountered in walks in the backcountry West. Stoltz's voice is a warm, slurring bass vibrato; set to Dobro, fiddle and female harmony vocals, it is nurturing and uplifting. Like John Prine, Stoltz is interested in the involvement of ordinary people in the history of a place, particularly where he feels the spirit of that place has been irretrievably lost. 'The Ballad of Willie and Millie' is about a couple who meet at a dance in the Columbia Gardens in Butte, Montana in the 1930s, before it was burned down and strip-mined in the early 1970s. Stoltz adds ruefully in his liner notes to *Listen to the Earth* (1989): 'The motherlode the mining company expected to find was never located'.

Stoltz's environmental advocacy emerges from a sense of nationalist pride, based on the myth of the American West as a scene of future potentiality. Yet his songs celebrate not the rugged individualism of the solitary man-in-nature, but sociability and generosity of spirit. As his notes on *Spirit Is Still On The Run* (1984) indicate: 'It's still out there. You can hear it in the mountain wind, feel it in the stillness of the desert, and taste it in the freshness of the wild places still left. Get out and experience it. Take a friend. The Spirit will never die as long as there are those who love and respect the Earth'.

Since the 1970s, under the continuing influence of Bob Dylan, many folk singers have aspired to be rock musicians, and 'going electric' has seemed the best option, both artistically and commercially, for singers such as John Prine and Neil Young, who began their careers in the acoustic folk tradition. But acoustic music, and the intimacy of the singer-songwriter mode that goes with it, underwent a revival with the emergence of 'Americana' as a radio format in 1995, a development which has led to the increased visibility of environmental protest folk music, in the work of Tish Hinojosa, David Rovics and Alice Di Micele, amongst many others.

In live performance, the Texan folksinger Tish Hinojosa dedicates 'Something in the Rain' (1990) to the Chicano militant labour leader Cesar Chavez, founder of the organisation that eventually became the United Farm Workers union. The song is told from the point of view of a young boy who watches his younger sister die, and comes to realise that his family's poverty and his sister's death are ultimately caused, not by the inevitability of nature, but by 'the sins of man whose profits rape the land'. The crop-spraying airplanes, which the boy used to believe 'cure the plants so things can grow', have been poisoning the land with pesticides. In Hinojosa's hands, the topical song has a narrative conciseness and psychological depth reminiscent of Woody Guthrie, while her measured guitar-picking and tender vocals add emotionality to political intelligence.

David Rovics is a prolific contemporary songwriter whose grasp of the drama and pithiness of the topical song recalls Phil Ochs. He puts the democratic, anti-capitalist values of the left-wing folk tradition into practice by making his songs available to download for free on the Internet. 'Here at the End of the World' (2004) illustrates well his revival of the art of plain speaking. A demotic, unaffected voice runs through a familiar list of environmental and political horrors: rising sea levels, the soil erosion and pollution caused by industrial agriculture and globalised trade, and the underlying cause of it all, class inequality. Rovics animates his political language with rhetoric familiar to the protest song: the intimacy of direct address, the turning around of colloquial expressions ('Let them eat coffee / Sugar, coca and flowers'), playful alliteration ('political pundits and corporate crooks'), and imagistic and metonymic details ('they sit in their mansions / On their plush leather chairs'). Musically, he builds up dramatic tension by varying the dynamics of his simple guitar strumming, based around a repeated minor chord, and releases it at the end of each refrain with a descending riff. In David Rovics, the tradition of folk music as environmental protest lives on, continuing the work begun by Pete Seeger and Malvina Reynolds in the early 1960s.

Alice Di Micele's' album *Too Controversial* (1990) positions the singer as a radical outsider, 'singing for survival' to an indifferent world. Among songs about poverty in America, President Bush's policies in Central America and apartheid in South Africa, are two songs about environmental concerns, which reveal the aesthetic strengths and weaknesses of the protest folk tradition. 'I Won't Say Goodbye' (1990) laments the logging of ancient redwoods. Di Micele meets 'a man in Philadelphia' who tells her of his lifelong ambition to visit the forests. She contrasts this with her reading 'in the newspaper something my senator said', about rather having logging jobs than old growth trees, and 'how he wished all the spotted owls were dead'. Taken as a whole, the song tends to lack those elements that make the topical song come alive in the hands of other singer-songwriters we have looked at in this chapter: John Prine's concrete sense of place and detail, Malvina Reynolds's political intelligence, Pete Seeger's sense of drama. Voice and music lack the nuances of Tom Paxton or the grittiness and energy of David Rovics.

In contrast, Di Micele's 'If I Was a Wolf' is more successful. The lyrics tell of the pursuit of a family of wolves by the Animal Damage Control and rangers from the US Fish and Wildlife service. The lyrics name specific places (Marion, Alberton), while soft rock drumming, power chords and tempo changes add a sense of drama to the narrative. The revenge fantasy of shooting the rangers with 'sleep inducing darts' is also provocative, possibly even 'too controversial', in that Di Micele felt the need to add a disclaimer on the lyric sheet: 'I was happy to hear that some local ranchers and Fish & Wildlife Service employees were sad and angry to see [the wolf] go. If I was a wolf, I'd chase those stereotypes.' The opinionated nature of the song at least means that it cannot be dismissed as what Paul Nelson called 'fish in the barrel stuff'.

True to the folk tradition of proletarian realism, the overriding mood of Tracy Chapman's album *New Beginning* is uplifting and optimistic. In the title track, the world is 'broken into fragments and pieces / that were once joined together in a unified whole'. Nevertheless, the song promises that 'we' can come together in a new beginning, and start all over. The reggae beat and rousing timbales shows a fusion of folk and world music styles, invoking the mood of 'One World' communality which will be explored in Chapter 10.

The most explicit environmentalist song on the album is 'Rape of the World' (1995), a bluesy lament made soulful by congas and mellifluous fretless bass. The lyrics list the familiar ways in which Gaia, the 'mother of us all', has been raped: clear-cutting, global warming, bombs 'exploded underground', and 'cities of concrete and steel'. The coda is a call to action: 'if you care you will stand and testify' to stop the rape of the world. Chapman's vocal performance enlivens these commonplace sentiments: she harmonises with herself through double-tracking, stretching syllables, breaking others off abruptly and adding spoken words. When she pauses slightly in her delivery of 'You seen her … stripped-mined', she breathes a sense of outrage into the line.

According to the aesthetics of political modernism, protest music is politically and aesthetically flawed. In the 1930s, Theodor Adorno argued that the German composer Hanns Eisler's attempts to forge a sense of community through class-conscious proletarian music were misguided, because he was merely trying to appeal to proletarian class-consciousness in its present form, which was degraded by capitalism. The aesthetic criteria valued by Eisler, namely, 'singability, simplicity, collective effectiveness *per se*', are, Adorno wrote, 'necessarily dependent upon a state of consciousness suppressed and enchained through class domination', and result merely in 'fetters (being) placed upon musically productive forces' (2002: 410). In other words, composers of proletarian music, by seeking to make music accessible to a mass audience, stretch neither their listeners nor themselves as composers, but instead merely perpetuate the conservative artistic values acceptable to the capitalist marketplace. Moreover, in capitalist society, according to Adorno, these 'communal and use' musics could only fail politically, because 'the human state of togetherness from which they proceed is a fiction and, where it might be something real, it is impotent when confronted by the capitalist process of production' (410).

While Adorno has an important point about aesthetic conservatism, his pessimism about folk music limits its aesthetic appreciation. The folk singers discussed in this chapter have produced work which, in various degrees, is musically and lyrically nuanced. It is possible too that environmental protest music has had a positive effect on the formation of ecologically aware individuals and communities in ways that should not be underestimated or dismissed out of hand.

## 6.3 Folk music and the 'American Primitive'

An alternative folk tradition in American music is less interested in overt environmental protest than the topical songwriters discussed so far, but nevertheless is deeply concerned with relationships with the natural world. Bob Dylan is again central to this tradition. At a press conference in Austin, Texas in September 1965, three months after his appearance at the Newport festival with a rock band, he distinguished between two types of folk music, or what he called 'historical-traditional music'. The main body of the music is not 'Depression songs', he said. Rather, it is 'based on myth and the Bible and plague and famine and all kinds of things like that which are nothing but mystery and you can see it in all the songs. Roses growing right up out of people's hearts and naked cats in bed with spears growing right out of their backs and seven years of this and eight years of that and it's all really something that nobody can really touch'. Folk music, he added, provides 'the only true, valid death you can feel today off a record player' (Marcus 1997: 30, 114).

The two folk traditions discussed by Dylan represent different attitudes to the natural world. In the 'mystery' tradition, nature is sublime and uncanny, an enchanted world that includes miraculous transformations between human beings and other living organisms. Dylan's reference to roses growing out of people's hearts is to the traditional Child ballad 'Barbara Allen'. In his own acoustic folk songs, Dylan evoked a similar sense of a world poised ambiguously and frighteningly between the natural and the supernatural. 'Lay Down Your Weary Tune' (1964), officially released on *Biograph* in 1985, is a pastoral rhapsody for a natural world experienced as music 'no voice can hope to hum'. Utopian yearning for individual freedom, for 'clouds unbound by laws', meshes with a sense of nature as cyclical as the structure of the song itself. In 'Girl Of the North Country' (1963), which Dylan based on the traditional British folk song 'Scarborough Fair', the harshness of the winter landscape, where the winds 'hit heavy on the borderline', is a correlative of the narrator's lost love. What unites both songs is a recognition of natural inevitability, which leads to acceptance of, and humility before, the powerful forces of external nature.

Harry Smith's 1952 *Anthology of American Folk Music* remains the most influential exploration of this dark and irrational element in American folk music. The ballads in Smith's collection, in the words of Greil Marcus, dramatise 'action; passivity; regret; sardonicism; absurdity; fear; acceptance; isolation; the wish for mastery running up against forces no one can understand, let alone master' (Marcus 1997: 107). Marcus calls this tradition the 'Old Weird America'. In contrast, the protest tradition in folk music has shown more confidence in conquering and mastering nature. In the environmentalist songs of

Malvina Reynolds and Pete Seeger, as noted earlier in this chapter, nature tends to be viewed as a resource for human use, although songs such as Peter LaFarge's 'Coyote, My Little Brother' suggested a biocentric critique of these human desires to master nature.

Marcus observes that the difference between the two folk traditions can be clearly seen in the changing artwork of the *Anthology of American Folk Music*. Harry Smith's original artwork invoked the late medieval astrologer and alchemist Robert Fludd, and included a sketch of what Smith called the 'Celestial Monochord'. However, when Marxist Irwin Silber took over the marketing of Folkways Records in the early 1960s, he changed the cover to a photograph of a hungry farmer taken by Ben Shahn in the 1930s for the Farm Security Administration, thereby replacing Smith's mystical allegory with proletarian realist protest. For Silber, the album protested the national shame of Appalachian poverty (Marcus 1997: 92-3).

Yet the differences between the 'mystery' and 'protest' traditions are not as mutually exclusive as Marcus, or indeed Dylan, suggest. Both can coexist in the same song. In Dylan's own protest songs, nature is often an untamed, mysterious force in the way Marcus outlines above. In 'The Ballad of Hollis Brown' (1963), for example, even the natural world appears to turn against the desperate, suicidal man, unemployed with a wife and children to feed. 'Bad blood' has taken his mare; his wife's screams stab him 'like the dirty drivin' rain'. When Hollis Brown takes a gun and shoots his family and then himself, a 'cold coyote calls' in the wilderness. The natural world appears to run according to a mysterious order unfathomable to human understanding: after seven people die on this South Dakota farm, 'somewhere in the distance' seven babies are born. The universe is made up of such uncanny correspondences beyond rational understanding, recalling Dylan's comment that the mystery tradition is about 'seven years of this and eight years of that'. Nevertheless, 'The Ballad of Hollis Brown' is also a powerful protest against the psychological effects of rural poverty and unemployment. What is absent from the song, though, is a sense of human agency: an individual or an institution that would be the object of one of Dylan's earlier 'finger-pointing' songs. In Malvina Reynolds's topical songs, as we saw earlier, she blames environmental problems on the California state government, absent landlords and the city fathers of San Francisco. In 'The Ballad of Hollis Brown', in contrast, no-one appears responsible, and therefore human agency also disappears: the overall mood is of resignation rather than of activist defiance. In this sense, the distinction between the two folk traditions, mystery and protest, is a valid one.

In the late 1990s, what Dylan called the 'mystery' tradition in American folk and blues music underwent a revival of interest, following the re-release of Harry Smith's anthology and Revenant Records' 'American Primitive' series of archive blues and gospel material curated by guitarist John Fahey. In the 1960s, Fahey had scorned the protest songs of the Folk Revival in favour of folk and blues performed by his alter ego Blind Joe Death (Pouncey 1998: 22-31). Fahey and the 'American Primitive' recordings became an important influence on the 'freak' or 'psych' folk movement which began in the early 2000s. In the songs of Devendra Banhart and Joanna Newsom, everyday perceptions of the natural world become playfully deranged, as they search for the Romantic primitivism of the child of nature.

Banhart's 'Little Yellow Spider', on his 2004 album *Nino Roja*, lists his imaginative encounters with nature's variety and abundance: yellow spider, white monkey, 'dancing' crab, snapping turtle, baby crow, 'sexy' pig, mockingbird, albatross and 'happy' squid. The narrator's attitude is fascinated and respectful: maybe the 'laughing' spider 'knows something that I don't know'. Banhart's biocentric respect for the otherness of animals recalls Captain Beefheart, whose work will be explored at length in the next chapter. Like Beefheart, he plays trickster and holy fool, his mission to rejoice in the wonders of a re-enchanted nature, and implicitly, therefore, to reduce his desire to interfere in it.

Joanna Newsom's songs explore a darker, more introspective world than that of Banhart, as she uses the imagery of the natural world, fairy tales and everyday domesticity to work through fears of loneliness and desertion. In 'Sprout and Bean' (2005), 'the hollow chatter of the talking of the tadpoles' suggests a sense of nature as impenetrable otherness which makes going outside both dangerous and alluring. 'Clam, Crab, Cockle, Cowrie' also sees animals and plants as imaginative objects into which a range of desires are projected: 'feels so good to be a rose', 'some great bellies ache / with many bumblebees'. Newsom's child-like voice moves from edgy and panicky to charming and fey, while her arrangements for harp subtly derange expectations of classical order and good taste.

If the natural world embraced by Joanna Newsom is that of the domestic garden and beach, for Ben Chasney's Six Organs of Admittance it is the Californian wilderness. *Dust and Chimes* (1998) sets finger-picked acoustic guitar in open tunings, reminiscent of the modal playing of John Fahey, against wistful, Oriental-inflected melodies and hand percussion. 'Dance Among The Waiting' suggests a pagan ritual in the woods at night: the repeated vocal line gradually becomes the howl of a wolf, and the droning guitar a waiting for mystical revelation.

The embrace of the natural world by the new, psychedelic folk music may lack the sense of activism and conservationist advocacy of the protest folk tradition, but in its more idiosyncratic and visionary relationship with nature, it constitutes a powerful alternative aesthetic to that tradition, and a revival of sixties countercultural utopianism comparable to that of the rave culture of the early 1990s, which we will examine in Chapter 11. The range of such intelligent and emotionally sensitive responses to the natural world, from Pete Seeger to Six Organs of Admittance, demonstrates the continued vibrancy of American folk music traditions.

# Chapter Seven

# 1960s rock and R'n'B

The 1960s counterculture was made up of two distinct groups, writes John Orman. What he calls the 'dope culture' used rock music simply for 'lighting up, boogying, making love, and relieving boredom'. The 'rock culture', on the other hand, took the music more seriously, considering it, in his words, 'the mechanism that would help form a new, alternative culture that would go "back to the land", live communally, "do dope", and try new sexual relationships while ignoring the establishment' (Orman 1984: 158). For this faction, rock musicians led the rejection of conservative ideas regarding sexuality, family, work and consumption; what the Jefferson Airplane's Paul Kantner summed up in the song 'Hijack', on his album *Blows Against the Empire* (1970), as the demand for 'free minds, free dope, free bodies, free music'. Expressed in this way, the preoccupations of the counterculture were largely humanist and anthropocentric. Yet allied to these concerns was a mystical view of the natural world that bordered on pantheism, shared by many musicians, with the notable exception, as we shall see, of Frank Zappa. 'Flower power' was a Romantic desire to return to what was perceived as a simpler, more natural form of social life based on the communal patterns of premodern tribal societies. The rock music associated with the hippie counterculture was integral to this newly emerging subculture of New Age environmentalism.

Moreover, several of the leading figures in the sixties rock scene went on to become environmental activists in the decades that followed. Joe McDonald became involved in animal rights and whale conservation; after two decades of environmental activism in his own locality, Ed Sanders of The Fugs founded the Woodstock Journal in 1995 with his wife, the writer and painter Miriam R. Sanders; Stewart Brand, former member of Ken Kesey's Merry Pranksters, who put on a taped music and slide show called 'America Needs Indians Sensorium' at the Trips Festival in San Francisco in January 1966, founded the

*Whole Earth Catalogue* in the fall of 1968, and in 1996 became a founder member of the Long Now Foundation, dedicated to promoting long-term thinking about the future of global ecosystems (Brand 1999).

Yet the nascent environmentalist critique produced by the rock counterculture was limited and contradictory, in that rock culture also stood for hedonism, individualism, egocentricity, escapism and consumerist fashion: values largely antithetical to radical environmentalism. Moreover, its adherence to a pastoral ideal of nature mystified questions of technology and economics, while its commitment to idealist notions of social change, particularly shaped by a naive faith in the transformative powers of both rock music and LSD, prevented the emergence of a more trenchant, materialist understanding of American society. In addition, the counterculture's inherent elitism, evident in its stress on youth as the new agents of revolution, and its consequent polarisation of society into the 'hip' and the 'square', prevented its members from addressing their own complicity in consumer society, and took the place of a more potentially radical questioning of class and racial division in the United States. It also made youth culture easily assimilable into the conservative discourses of the mass media (Cantwell 1996: 307-8; Frank 1997: 14). Consequently, in the 1960s, as Eric Mottram put it, the 'energy of the searching young in America remained depoliticised and exploited' (1988: 215).

Rock songs, writes Robert Pattison, 'are either about the ideal triumph of self as it moves to encompass more territory or about the splenetic self-questioning that accompanies failure of this goal' (1987: 110). In the 'secular pantheism' of the rock performer, it is as if the self expands to embrace the whole universe. According to the Romantic anarchist politics of the counterculture, lifting social repressions would liberate the essential creativity and goodness of this autonomous, individual self.

For poet Allen Ginsberg, the cosmic self apparently revealed by LSD prefigured a new ecological relationship with one's environment. Most 'acid' mysticism was vague and egocentric, typified by the admonishment to 'feed your head' in the Jefferson Airplane's 'White Rabbit' (1967a). That LSD culture focused on such an Alice-in-Wonderland trips of individual self-discovery, rather than on the formation of a wider social and ecological self, was a problem that Ginsberg strove to address in his writings in this period.

He and fellow poet Gary Snyder discussed ecological consciousness after the Human Be-In, or Gathering of the Tribes, held in San Francisco in January 1967. Ginsberg understood that the growing sense of mystical unity should include ecological awareness, and that rock music would play a vital role in its propagation. As the title of his 1968 essay 'Consciousness and Practical Action' suggests, he attempted to merge Buddhism with Communism by positing a self that could expand, through the mediation of LSD, towards a sense of care for the natural world. Ginsberg described the oceanic feeling of an LSD trip as a 'unitive experience of One, of all of us being one – not only ourselves with varying colour of skin and mysterious ego-origin or whatever we are, also one with flowers, also the very trees and plants'. For Ginsberg, the Rolling Stones and the Beatles embodied what he called 'slow communalisation', a social process which was 'the centre of consciousness and also the centre out of which political activity begins'. What was emerging, and what he tentatively attempted to describe, was the 'beginning of a friendly communism, or communion, or community, or friendly extension of self outward [...]' (Berke 1969: 173).

In 1967, the *San Francisco Oracle* published 'Ten commandments for getting to the city of God', one of which was to 'Revere nature'. In the same year, the underground newspaper also rewrote the Declaration of Independence: 'We hold these experiences to be self-evident, that all is equal, that the creation endows us with certain inalienable rights, that among them are: the freedom of the body, the pursuit of joy, and the expansion of consciousness' (Miller 1991: 109). The rock culture had thus begun to take the difficult steps towards the biocentric implications of the idea that 'all' (not just 'all men') is 'equal'.

Evan Eisenberg writes that, in the sixties, rock music 'moved outwards to embrace all life and preserve it in vinyl' (1987: 80). Yet vinyl records are, of course, a product of the oil industry: the counterculture's dream of unity with the natural world through rock music thus depended on the very technologies that were threatening that world. How it dealt with such contradictions is the subject of this chapter.

## 7.1 Sixties rock music as electronic pastoral

Sixties rock music was an electronic pastoral, in which a desire to go back to nature, both rural and wild, interacted, unstably but creatively, with an apparently contradictory embrace of the American technological sublime (Nye 1994). As musicologist David Schwarz observes, rock music is, in many respects, a modernist discourse, given 'the increasing glorification of technology in musical instruments and sound systems that reproduce rock and roll, the urge to find increasingly "new" sounds, forms of expression that surpass antecedents, the massive, public spectacle, an increasing monumentality of compositions, and glorification and commodification of individual subjectivity' (1997: 98). Yet the obverse of this technological modernism is a fascination with the archaic and primitive. Pattison comments that this 'apparent paradox' is in fact 'two flanks of a single movement, the leading spirit of which is selfhood militant'. He explains that, for the rock performer, 'Man's proper home is the pastoral Eden of accomplished feeling where self is identical with the universe […] One road to this leads through the primitive past of Romantic mythology, but another lies directly through scientific invention. Both converge on infinity' (1987: 126).

Rock music has always blithely integrated nature and technology in this way. After all, one of its founding icons, Chuck Berry's Johnny B. Goode, was a country boy, born in a cabin 'made of earth and wood' in rural Louisiana, who learned to play guitar by 'strumming with the rhythm' of the passing railroad trains. But when rock music challenged the Folk Revival in the mid-sixties, many of the new rock musicians were drawn to metaphors of more contemporary technology, particularly jet aircraft, space rocketry and nuclear energy, to describe the 'far out' explorations of their amplified sound. In so doing, they appropriated modern technology as both a metonym of social progress and a metaphor for their quasi-mystical explorations of the inner self. Steppenwolf's 'Born to Be Wild' (1968) encapsulated this conflation of nature and modern technology, whereby to be 'like a true nature child' is to identify oneself with technological speed and power ('Get your motor running'). The use of the song on the soundtrack of the movie *Easy Rider* (1969), accompanying images of motorcycles opening up the vast spaces of the American West, vividly celebrated the expansion of this frontier-crossing, masculine self, taking the whole world in a 'love embrace'.

In *The Electric Kool-Aid Acid Test*, Tom Wolfe described Ken Kesey, leader of the Merry Pranksters, as the product of the 'technological superheroics of the jet, TV, atomic subs, ultrasonics – Postwar America suburbs – glorious world! and the hell with the intellectual badmouthers of America's tailfin civilisation' (1969: 35). Ironically, then, the counterculture was complicit in the very industrial society its rhetoric rejected. Ultimately, the rock counterculture's fascination with the technological sublime was guaranteed by the cornucopian optimism of sixties 'post-scarcity' economics, a leading advocate of which was the inventor R. Buckminster Fuller, who, as we saw in the previous chapter, became an important authority for figures as diverse as John Cage and John Denver. The optimism of post-scarcity economics would later be challenged by the neo-Malthusian counter-discourse of austerity and limits that emerged in the early 1970s. In the sixties, however, the widespread belief that material scarcity had already been transcended by technological know-how provided for the rock culture a way of ignoring questions about the environmental costs of the global electronic technologies it embraced.

Rock music fetishised its electronic technologies: the electric guitar, amplifier, microphone, magnetic tape and television. Eric Mottram makes the connection between the rock musician and that other icon of 1960s technological pioneering, the astronaut: 'The electro-visual aspects of performance – wires, microphones and speakers – gear the human body into the energy of circuitry as much as an astronaut becomes the cyborg of his capsule' (1988: 183). Captain Beefheart told an interviewer: 'Electricity never stops moving. That's the 'spark of life.' How can you turn life on or off?' (Chorush 1971: 4).

That an atavistic sense of cosmic Oneness could be gained through advanced electronic technology was a contradiction enthusiastically embraced by the rock culture. That culture was an artistic product of what Marshall McLuhan called 'all-at-onceness': the effect that twentieth-century electronic technologies were having in making communications global in space and simultaneous in time (1962: 63). In June 1967, rock became a global phenomenon when the world's first live international satellite television broadcast featured the Beatles' 'All You Need Is Love', and was watched by an estimated audience of 700 million people (Lee and Shlain 1985: 179). In 1969, Paul Kantner of the Jefferson Airplane called the loose group of musicians with whom he was collaborating the Planet Earth Rock and Roll Orchestra. The mystical implications of such global connections were not lost on Allen Ginsberg, who spoke of wanting to make San Francisco 'an electric Tibet' (Lee and Shlain 1985: 176).

Musicians often evoked the technological sublime when trying to explain the creative origins of their music. In the sleeve notes to *Mr.*

*Tambourine Man* (1965), guitarist Roger McGuinn commented on The Byrds' electrification of folk music: 'The instrumentation is changing somewhat to meet the nuclear expansion and jet age'. Jet travel formed the literal interpretation of 'Eight Miles High', on *Fifth Dimension*, which the band favoured when some radio stations banned the song for its alleged subtext about hallucinogenic drugs. The song, McGuinn said, was about the dislocating perceptual effects that jet air travel had on the band when, as visiting Americans, they found themselves unable to understand the street signs in London. As McGuinn's twelve-string guitar adopts modal harmonies inspired by John Coltrane and the Indian sitar-player Ravi Shankar, these disorientations of space and time, an experience 'stranger than known', are associated with Eastern mysticism.

Jerry Garcia of The Grateful Dead also used the metaphor of a jet aircraft to describe the band's music, likening his guitar feedback to 'your mammoth bringing of a plane in for landing at the end of a huge jam'. The origin of the band's experiments with loud, immersive volume and unpitched noise lay in an experience they had in the hills above Palo Alto, when a jet aircraft flew low over their heads as it came in to land at San Francisco airport. The noise it made, Garcia said, 'just took everything, it took all of history with it. The sound was absolutely cataclysmic, a huge sound. Right there for me was a moment in which everything – all sound, music, and everything else included – was born again' (Troy 1994: 123-4). In Garcia's musical epiphany, rock music's extreme volume could evoke an apocalyptic sense of death and rebirth that transcended linear, historical time. In this way, immersive listening resembled the experiences of both psychedelic drugs and Eastern religions.

'Dark Star', The Grateful Dead's most famous extended improvisation, developed this exploration of immersive sound, loose musical structures and open-ended forms as a science fiction adventure. Garcia described the song, on the album *Live/Dead*, as 'exploratory ventures – possibly you could use the word experimental for that – it's not so much a set piece, that you know where you are in it and know where you're gonna go, as you're out on an ocean in a boat and you can choose your landmarks and response to things and move in certain directions as you wish – of course, always interacting' (Troy 1994: 123). As 'Dark Star' develops, meandering guitar solos eventually return to a central melodic motif, one of the 'landmarks' to which Garcia refers (Boone 1997: 171-209).

Garcia's description of the band's music as like a boat on the ocean combines metaphors from the natural world and technology. Some of his descriptions went further, and conflated modern technology with nature. 'There were times', he said, 'when the feedback passages at the end of a long jam were so powerful, listeners would swear it was thunder or the earth opening up' (Troy 1994: 123-4). Such a statement matched the nature metaphors favoured by the band's percussionist Mickey Hart, who in 1970 described the sound that he and fellow drummer Bill Kreutzmann tried to achieve as, 'Earth music ... The rhythm of the earth, like I get riding a horse' (Lydon 1972: 562). Yet for the sixties counterculture, the exploration of outer space was a particular focus for this fascination with both advanced technology and the natural world. The trope occurs not only in The Grateful Dead's 'Dark Star', but also in songs by The Byrds, Jefferson Airplane and Jimi Hendrix, amongst many others.

When the Apollo 11 astronauts landed on the moon in July 1969, Norman Mailer considered the event proof that the Squares had won. In *A Fire on the Moon* the bank of instruments in the space capsule represented for him 'that sense of life programmed and wasted, of reason so over-applied to life that all contradictions having been killed, the light of reason had finally left the eye [...]' (1970: 152). For Mailer, as for Thomas Pynchon, then writing his epic novel *Gravity's Rainbow* (1973), space rocketry was the frontier technology of the modern totalitarian state.

In the rock culture, Captain Beefheart stood out in his opposition to the moon landing, believing it to be a masculinist rape of nature. He told an interviewer that 'they're about to poke their genitals into our cream cheese moon right now', adding, 'That's my eye'. The moon, he said, is 'part of me' (Barnes 2000: 114). For Beefheart, as for the deep ecologists who were to come to the fore later in the 1970s, the universe is part of his cosmically expanded self, and therefore the subject of his ethical concern.

Beefheart's outright rejection of the space program was unusual for a rock musician. Bob Dylan was to make a similar indictment of male hubris on 1983's *Infidels*, when in 'License to Kill', still under the influence of his recent conversion to Christianity, he indicted the space program as male permission to violate nature: 'man has invented his doom / First step was touching the moon' (Heylin 2000: 497-500). For Dylan, space exploration continues to an apocalyptic degree the same male destructiveness experienced locally by the 'woman on my block', sitting in fear of a more domestic form of violence.

In the 1960s, however, outer space travel often featured more positively in rock music, as the counterculture mixed fascination with big technology with a desire to appropriate the imagery of space rocketry for its own ends, in defence of a wider notion of rationality than that which, as Mailer indicated, informed the Apollo project itself. The popularity of Stanley Kubrick's film *2001: A Space Odyssey* (1968) with hippie audiences was an example of this act of appropriation, as was Roger McGuinn's work with The Byrds.

The Byrds released a series of albums in the mid-sixties in which they brought together a pastoral view of nature with an embrace of futuristic modernity. In effect, the continued existence of the former guaranteed the beneficence of the latter. The two contradictory influences can be directly attributed to the clashing interests of Roger McGuinn, who wanted the band to be an electronic, jazz influenced group, and bassist Chris Hillman, who had played mandolin with his bluegrass band the Hillmen in the early 1960s, and wanted to steer the band towards country music.

McGuinn's '5D (Fifth Dimension)' is a futuristic evocation of cosmic consciousness, in which the space traveller opens his heart 'to the whole universe', and finds it to be 'loving'. The weird beauty of sensory displacement is invoked by the celestial-sounding organ part, for which McGuinn told Van Dyke Parks to 'think Bach'. McGuinn described the song as being about the existence of multiple dimensions beyond height, width, depth and time. 'A lot of our world is very materialistic and scientific', he wrote. 'It overlooks the beauty of the universe'. For McGuinn, then, a mystical view of the universe prevailed, and travel in outer space, rather than being a product of scientific rationality, revealed the limits of what the song called 'scientific delirium madness'.

The Byrds played off their embrace of high tech futurism against a paradoxical use of deliberate archaism, derived from Anglo-American folk traditions. '5D (Fifth Dimension)' placed the spatial and temporal paradoxes of space travel within the traditional form of a sea shanty. Some of the band's other science fiction songs also shared this stylistic trait. 'C.T.A-102', on *Younger than Yesterday* (1967), celebrated

American technological pioneering with an archaic turn of phrase: 'We're ready for to go / out into the universe'. The more portentous 'Space Odyssey', on *The Notorious Byrd Brothers* (1968), was based on Arthur C. Clarke's story 'The Sentinel', the same source that Kubrick later used for *2001*. Another sea shanty, the song describes 'evolving man' awaiting redemption from aliens beyond the solar system. Evolutionary progress has been made possible by 'our atomic energy', which 'we harnessed and survived'.

This use of folk archaisms took the curse off the band's technological futurism, including their endorsement of nuclear power, by framing it within reassuringly familiar discourses, a process which Marshall McLuhan called looking 'at the present through a rear-view mirror' (McLuhan and Fiore 1967: 74-5). Significantly, the giddy, destabilising modernism of '5D (Fifth Dimension)' was followed on the album by Francis McPeake's folk song 'Wild Mountain Thyme', which brought to mind a sentimental pastoral scene of love amongst the heather. For The Byrds, it was possible to have it both ways. Significantly, the arrangement of 'Wild Mountain Thyme' for string orchestra suggests urbanity and sophistication, and made no attempt to evoke the rural authenticity that Gram Parsons wanted for the band when he joined in 1968, and which, as we will see in the next chapter, typified the country rock era of the late sixties.

Historian David Nye notes that manned space flight appealed to the American public in the 1960s as a positive use of technology, in contrast to the apocalyptic fears that surrounded the nuclear arms race. The rock culture shared this view, as can be seen in the songs against nuclear weaponry which rock adopted from the protest folk tradition, the most successful of which, in commercial terms, was Barry McGuire's 1965 hit 'Eve of Destruction'. The Byrds' 'I Come and Stand at Your Door' (1966), on *Fifth Dimension*, was based on Pete Seeger's version of the poem by the Turkish poet Nazim Hikmet about a seven-year-old child killed in Hiroshima. The Grateful Dead sang 'Morning Dew', a song about nuclear annihilation written by folksinger Bonnie Dobson, at the Human Be-In in January 1967, and released in that year on their eponymous debut album. The cover of the Jefferson Airplane's *Crown of Creation* (1968) showed the mushroom cloud from an atomic bomb, while 'House at Pooneil Corner' evoked a strangely deserted and scarred Earth devastated by a nuclear holocaust.

That the technologies of nuclear weaponry, nuclear power and space flight were all integral to the American military-industrial complex could easily be forgotten in the sublimity of an Apollo rocket launch. 'In an atomic age', writes Nye, 'the pilgrimage to the Kennedy Space Centre promised a sublime experience that renewed faith in America and in the ultimate beneficence of advanced industrialisation. This final avatar of the technological sublime is a literal escape from the threatened life-world' (1994: 256).

This notion of space travel as an escape from an ecologically damaged world was taken up in a suite of interconnected songs on the second side of Paul Kantner's 1970 solo album *Blows Against the Empire*. In Kantner's utopian science fiction fantasy, the hippie counterculture escape from the Earth in a hijacked spaceship, leaving behind two impending catastrophes: global overpopulation and civil war in America after the state repressions of 1968. 'Hijack' is a paean to the utopian promise of the electronic pastoral: 'The land is green and you make it grow, go to the forest and move'. This appeal to wild nature is associated with the 'thundering electrical energy' of an outdoor rock concert in a park. The song's drifting, additive structure and improvised interplay enacts the sense of rock music as explorative travel.

In an interview with *Rolling Stone* journalist Ben Fong-Torres, Kantner described how his science fiction fantasy was inspired by Robert Heinlein's novel *Stranger in a Strange Land* (1961), which, with its celebration of 'free love', communal living and mystical interconnectedness (what the novel calls 'grokking'), was a big success with the counterculture. The song 'Starship', Kantner said, is 'about us – me and Jerry Garcia and David Crosby stealing a starship – hijacking a spaceship, going where whoever comes along wants to go. It's my answer to the ecology problem. It's the only way it's all going to get together and work' (Fong-Torres 1970: np). That Kantner's starship project was to be researched by Owsley Stanley III, the millionaire LSD chemist and sound technician for The Grateful Dead, suggested an allegorical reading of the album in terms of the liberatory possibilities of exploring 'inner', as well as 'outer', space.

Kantner's futuristic utopia is both pastoral and urban ('our babes'll wander naked thru the cities of the universe'), and attempted to reconcile an ecologically benign natural world with modern, futuristic technology. The image of piloting a spaceship recalls Buckminster Fuller's contemporaneous reference to Spaceship Earth, which posited the ecological health of the planet as an issue of technocratic control (Fuller 1969). As a rock and roll star, Paul Kantner's faith in the radical potential of enlightened technocracy and post-scarcity technological pioneering is guaranteed by its complicity in the American mythologies of the good outlaw and the open road. Deliverance for the counterculture will come through the synergy of his favoured technologies: the recording studio, cinema, space rocketry and LSD.

As with all utopias, then, the text of the future was fissured by the contradictions and conflicts of the present. Kantner told Fong-Torres: 'Unless we have a war or a big disease or a famine, there's just too many people, and they're gonna have to get off the planet. This is my way of starting to get off a little earlier' (Fong-Torres 1970: np). Salvation is reserved for the select few, as Kantner reinforces the elitist dichotomy between the 'hip' and the 'square', with its idealistic faith in the innocent promise of youth, that, in hindsight, seriously limited the politics of the sixties counterculture.

A similar sense of moneyed privilege and political complacency can be seen in David Crosby's 'Wooden Ships', which he wrote with his friend Kantner on his sailing boat in June 1969. (Different versions of the song appeared on Jefferson Airplane's *Volunteers* and Crosby, Stills and Nash's eponymous 1969 album). Sailing in a yacht on a pastoral ocean provides refuge for the nouveau riche rock stars from the impending collapse of American society. The privileged youth leave the 'silver people' behind on the shoreline, in an act of anti-social withdrawal: 'All we can do is echo your anguished cries / Stare as all human feelings die'. Departing from the binary verse-chorus structure typical of rock music, the song depicts a journey of deliverance from the ordinary and the mundane. According to Fong-Torres, David Crosby and Dino Valenti of the Quicksilver Messenger Service 'talked about having a boat and being able to split at any time, as an alternative to fighting in the streets'. 'Wooden Ships', Barney Hoskyns notes, 'articulated the escapist elitism of the new Laurel Canyon aristocracy' (1996: 129). Its survivalist mentality was also an important aspect in the back-to-nature movement emerging in the counterculture at this time.

In 'After the Goldrush' (1970), Neil Young used similar science fiction tropes to Kantner, depicting an ecologically damaged world but without his elitist utopianism. Mother Nature is 'on the run in the nineteen seventies', while, in the final verse, Young dreams that the 'chosen ones' are preparing silver spaceships to fly 'Mother Nature's silver seed to a new home in the sun'. In Young's dystopian near-future, though, elitist escape from a doomed Earth is not an option; he is left instead 'lying in a burned-out basement'.

The electronic pastoral embraced by the sixties musicians discussed so far in this chapter was largely an anthropocentric concern: the ecstatic interaction of nature and technology in the music of The Grateful Dead, The Byrds and Paul Kantner served to expand the potentialities of the self into a community defined mostly in human terms. However, psychedelic consciousness was also moving in the direction of a more biocentric sensibility, as the following pages will explore.

## 7.2 Biocentrism in sixties rock music

The idea that the natural word is sentient, and therefore worthy of ecological concern, pervades the rock music of the late 1960s. Jefferson Airplane's 'Eskimo Blue Day' (1968) indicted human arrogance in the face of a sublime nature. The redwoods 'talk to me', sings Grace Slick, and their message is that human ambition 'doesn't mean shit to a tree'. Slick commented on her song: 'Our greed *does* mean shit to a tree. The trees are dying. All of our separating ourselves from the planet is stupid because, the larger picture, whether or not you become president of Bank America has nothing to do with evolution' (Tamarkin 2003: 204).

In The Byrds' 'Dolphin's Smile' (1968), on *The Notorious Byrd Brothers*, wild animals embody a childhood dream of floating aimlessly on the ocean, innocent and free. The introductory bars mimic dolphin calls by adding electronic echo to the sound of Roger McGuinn's fingernails tapping on the neck of his guitar. The bridge passage evokes the strangeness of the dolphins' world, as cosmic consciousness edges into a fantasy of biocentric egalitarianism. The Beach Boys' 'A Day in the Life of a Tree', on *Surf's Up* (1970), is a slow dirge written from the point of view of a tree dying because of air pollution: 'the pain, the air is killing me'. Bird song effects are layered over sustained chords on a child's toy organ to connote the innocence of nature betrayed by human greed.

But it was the music of Captain Beefheart which explored biocentric awareness in a sustained manner. 'People are just too far out', he told Langdon Winner. 'Do you know what I mean? Too far out – far away from nature' (Winner 1971: np). Beefheart's lyrics present a vision of the natural world familiar to deep ecology. Left free of human intervention, nature would be a benign arcadia. Ecological redemption will come when human beings reconnect with wild nature, both outside and inside themselves. For Beefheart, industrial technology ends in death, as he sings in 'The Dust Blows Forward 'n the Dust Blows Back', on *Trout Mask Replica*: 'the smokestack blows up in sun's eye / When am I gonna die?' He is tempted by rural nostalgia as an escape route: in 'She's Too Much For My Mirror', he longs to leave behind the corruptions of Chicago and return to that 'little red farm' in 'Lousey', with its butterflies and cornfields. However, Beefheart also questions the very possibility of the arcadian idyll he desires. In 'Clear Spot', the title track of his 1973 album, it is the natural world that he wants to run away from, as he finds himself in a rotting, stinking bayou threatened by mosquitoes and a burning sun.

'Steal Softly Through Snow', on *Trout Mask Replica*, expresses a desire for unity with the natural world, while acknowledging an inevitable sense of distance from the desired object, which registers as an emotional lack. 'The moon frayed thru dark velvet lightly apart', Beefheart sings: the displacement of 'frayed' from the dark velvet curtains to the moon itself registers his disappointment at not gaining direct access to nature, so that the moon itself appears spoiled. This disappointment is doubled in the following line, when, broken-hearted, he is unable to follow the winter migration of the wild geese.

Later in the song, when Beefheart sings the line 'Grain grows rainbows up straw hill', the authenticity of the pastoral idyll is undermined by the arch, melodramatic vibrato in his voice. Although Beefheart largely embraced the ruralism that many hippies advanced as an alternative to the war society, his performance thus tempered it with irony. The next line explains why, as, in the polemical climax of the song, industrial development and war are revealed, in Leo Marx's terms, as a counterforce that interrupt the pastoral idyll: the highway across the hill breaks his heart, while 'Man' has 'lived a million years 'n still he kills'. As Beefheart's vocal delivery becomes harsher, drummer John French hits the cymbals to reinforce the anger. For Beefheart, militarism and industrial development are both part of a single war against the natural world. In this context, the song's refrain – 'steal softly thru snow' – recalls the emerging radical environmentalist call to 'tread lightly upon the earth' by placing minimum demands on the Earth's ecology.

Included in Beefheart's interest in the natural world is a fascination with everyday technological transformations of it. The very title *Trout Mask Replica* connotes two layers of distancing from nature. The photograph on the back cover shows the band, not against a backdrop of pristine nature or old Americana favoured by his rock contemporaries, but in a suburban back garden in Los Angeles. Beefheart is holding a lampshade. His lyrics similarly complicate his desire for Romantic retreat by using recurring images that conflate nature with modern technology. In 'Orange Claw Hammer', an oriole flies into the air 'like a bomber', its tail clawing the evening 'like a hammer'; in 'Well', black birds fly with 'feathers of solid chrome'. The title 'Neon Meate Dream of a Octafish' conflates modern technology, wild animals, sexuality and the unconscious, the faux-medieval spelling suggesting magical transformations of human into animal.

Beefheart's prankster interest in role-playing and imaginative transformation implies that human identity is multiple and dynamic, just as his voice and music are themselves protean in form. As he commented on his music in 1991: 'It is trying to break up the mind in many different directions, causing them not to be able to fixate [...]' (Barnes 2000: 91). Beefheart's conception of a cosmically expansive, unstable self grew out of psychedelic culture. However, having released 'Ah Feel Like Ahcid' on *Strictly Personal* in 1968, he later took steps to distance his work from psychedelic rock, instead locating the source of his visionary experience in the natural world, rather than in LSD. He told *Zig Zag* magazine in 1969: 'A trout in a stream. If you have to say psychedelic, that's psychedelic to me' (Frame 1969: 3).

Intimate, close observations of the small sights and sounds of the natural world are thus central to Beefheart's vision. In 'The Dust Blows Forward 'n The Dust Blows Back', he notes a blue jay's beak 'open an inch above a creek', and can hear the sound of the 'mice toes scampering'. In 'The Buggy Boogie Woogie', on *Lick My Decals Off, Baby* (1970), he guiltily disturbs a 'momma spider' and her nest when cleaning a 'two-dollar room'. It is this ethical care for the natural world that most distinguished Beefheart from his rock contemporaries, and fulfilled Allen Ginsberg's call for an inclusive, ecological sense of self that we saw earlier in this chapter. 'All of *Trout Mask Replica'*, Beefheart commented in 1973, 'was about ecology' (McGrath 1973: 2).

In some respects, Beefheart's enactment of this ecological self prefigured the transpersonal self of deep ecology. In 'Frownland', the opening track on *Trout Mask Replica*, Beefheart defiantly announces that his spirit is made up of the ocean, the sky, the sun, the moon and 'all my eyes can see'. Yet this identification of the self with the rest of the universe is complicated by an appeal to a more individualistic notion of personal autonomy. With the guitars playing in a jerky, idiosyncratic 7/8 time signature, Beefheart declares: 'I want my own land'. Moreover, the archly comic, knowing tone of his vocals deflates the pretensions of his expanding self, thereby preventing what Geoff Ward, in reference to Ginsberg's poetry, calls the 'vatic ambition of Beat prophecy and self-absorption' (1993: 29). In Beefheart's lyrics, visionary pronouncement is never allowed to become transcendental, because it is ironically undercut by his recognition of mass culture in its intractable, banal materiality. Everyday material culture thus provides a check on the self's ambitions. In 'Debra Kadabra', on *Bongo Fury* (1975), he is smothered in Avon cologne, and forced to watch the Rubber Tongue cartoon on Channel 13. His spoken introduction to side two of *Trout Mask Replica* – 'A squid eating dough in a polyethylene bag is fast 'n' bulbous, got me?' – brings the natural into violent contact with the artificial, in a humorous image of confinement, consumption and monstrous excess.

Beefheart's ecological vision was based on an arcadian ideal of non-violence and co-operation. In 'Ant Man Bee', the bee is a model for this benign ecology, by setting the flower free when it takes its honey. In contrast, 'Man' and the ants 'won't set each other be'. The verbal puns, free-blowing horns and jagged, syncopated rhythms enact the joy that the bees promise for humankind, if only it were to follow their example. Referring to 'Ant Man Bee' in 1993, Beefheart commented: 'Horrible. Human beings' (Barnes 2000: 99). The sense that, for Beefheart, human nature is essentially corrupt is confirmed in the pessimistic title of another song on *Trout Mask Replica*: 'My Human Gets Me Blues'. Accordingly, Beefheart prefers animals to human beings. In 'Wild Life', the narrator intends to go up into the mountains, find a cave, and 'talk the bears / into takin' me in'. Wild life, he says, is 'a man's best friend'. The friendly animals in Beefheart's songs recall those in early twentieth-century nature writers such as Ernest Thompson Seton and William Long, and, as we will see in Chapter 9, formed the basis for what became in the 1970s an explicit concern for animal rights.

Parallel to Beefheart's celebration of arcadian nature was a protest against the threats posed by human beings to the health of the natural world. He remarked about the front cover photograph of *Trout Mask Replica*, in which he is seen wearing a carp head as a mask: 'What I was saying was that the carp seems to be able to thrive in polluted waters, and I'm waving to tell people that no-one else thrives on pollution' (McKnight 1972: 3). He variously described 'Safe As Milk' (1967) as being about both radiation and D.D.T. in mother's milk (Frame 1969: 1; Boucher 1972: 2).

By the early 1970s, such anxieties over environmental apocalypse began to have a dominant role in Beefheart's music. The title track of *Lick My Decals Off, Baby* (1970) takes up his familiar theme of a lost paradise, telling of the birds and bees 'and where it all went wrong'. This notion of a fall from an Edenic state of nature is taken up again in 'Petrified Forest', which warns that the human race is heading for extinction like the dinosaurs. The 'rug's wearing out that we walk on', Beefheart warns. It is being destroyed by those who 'breathe in 'n' don't breathe out', merely consuming the Earth rather than engaging in a mutual exchange with it.

In a *Rolling Stone* interview in 1971, Langdon Winner recounted a visit he made with Beefheart to the Smithsonian Institute in Washington DC, where he watched him become visibly angered by the display of stuffed birds in the 'Extinct Birds' section. Looking at the carrier pigeons, Beefheart said: 'I can't believe it. Look at that. It's paradise. Man had paradise and he blew it'. Before viewing the dinosaurs, he commented: 'We might as well go see how we're going to end up' (Winner 1971: np). Beefheart considered music a way of trying to break up the mental habits that he believed ultimately led to such abuse of the natural world. Music, from this Romantic point-of-view, was a way of returning to an unspoiled nature before the degradations introduced by human civilisation. As we will see in the following section, Beefheart's fears of environmental apocalypse were shared by many of his rock contemporaries, who came up with their own musical responses to it.

## 7.3 Environmental apocalypse in sixties rock music

Lawrence Buell writes that what he calls 'toxic discourse' emerges when disenchantment with the pastoral ideal of nature 'is accompanied or precipitated by totalising images of a world without refuge from toxic penetration' (2001: 38). A central trope of such discourse is 'the shock of awakened perception' (35). In the late sixties and early seventies, rock music was pervaded by this sense of a rude awakening to an impending apocalypse that extended to the death of the natural world itself. The sixties counterculture embraced what Gregg M. Campbell calls a tradition of 'green or pastoral' apocalypse, which sought to remind its Elders of 'the pastoral vision that had been the original inspiration of the American Dream' (1998: 98).

Rock critic David Downing notes that, in the 1960s, popular songs turned apocalypse into 'a fad and a cliché' (1976: 28). But the rock music of the sixties also made environmental apocalypse perversely attractive. 'Everywhere in Dylan and The Doors', writes Eric Mottram, 'the message is: this is The End. American resources have been exhausted. A prophetic sense that the function of the rock synthesis is to celebrate the apocalyptic and even holocaustic collapse of a culture impregnates the sounds and the words' (1988: 188-9). Environmentalist Paul Hawken writes that the concept of doomsday 'has always had a perverse appeal, waking us from our humdrum existence to the allure of a future harrowing drama' (1993: 204).

Part of the emotional appeal of apocalyptic traditions is in the way they encourage a sense of communal identity through the creation of a shared enemy (O'Leary 1994: 215-6). This sense of apocalypticism as an assertion of collective identity can be seen in the way The Fugs' Tuli Kupferberg welcomed impending social and environmental collapse as an opportunity for revolutionary upheaval. He wrote in 1968:

> When the wheels stop, when the lights go out, when the subways grind to a halt, when there's no fuel for the automobiles (or bombing-planes), when the teachers strike, when the garbage festoons (!) the streets, when you drug yourself, when the beautiful storm rages, impelling taxis to stop & caterpillars to cuddle: then my friend is when things begin to happen; then is when life begins! [...] The patterns break! Your old life-self is dying! Help help someone! Anything is better than my unknown! (Kupferberg 1969: 86).

Kupferberg's enthusiasm for social and environmental apocalypse, irrespective of the widespread suffering it would cause, has a similarly elitist sense of election as Paul Kantner's 'Hijack' and David Crosby's 'Wooden Ships', discussed earlier in this chapter.

Rock music provided both a soundtrack for the impending apocalypse and a therapeutic containment of the anxieties it generated. Journalist Lester Bangs wrote that the Rolling Stones 'were dirty, but The Doors were dread. [And] dread is the great fact of our time' (Hoskyns 1996: 154). But the dread evoked by The Doors' music was alluring and strangely reassuring, as protests against social and environmental collapse edged into celebration. 'When the Music's Over' (1967) became an anthem for the impending environmental apocalypse. In its spoken interlude, Jim Morrison asked melodramatically: 'What have they done to the Earth? What have they done to our fair sister?' The answer, delivered in staccato accents, is brutal stabbing and rape. The demand in the refrain – 'We want the world, and we want it ... NOW' – was adopted as an anthem by anti-war demonstrators, indicating, as Lee and Shlain observe, their dismissal of 'any notion of temperate or deliberate change' (1985: 216). Yet The Doors' music also provided therapeutic uplift. Music, Morrison tells his listener, 'is your only friend until the end'. Morrison's crooning voice and Ray Manzarek's funfair Vox organ made urban fear and alienation pretty and delirious rather than terrible. As writer Eve Babitz perceptively said of Morrison: 'He was pure Hollywood. It was showbiz meets Aleister Crowley' (Hoskyns 1996: 158).

The early 1970s saw a flurry of songs about environmental apocalypse, as the first Earth Day on 22nd April 1970 further brought environmental issues into the public arena, and signalled the adoption of policies of cautious environmental reform by government and big business (Anderson 1995: 348-50). *Earth Rot*, by jazz arranger and producer David Axelrod, was the most ambitious attempt to represent environmental apocalypse in popular music. Capitol Records planned to release this epic song suite on university campuses to coincide with the first Earth Day (Rudland 2001: np). The lyrics were adapted by Axelrod's brother Michael from the Old Testament Book of Isaiah and the Navajo 'Song of the Earth Spirit'. 'The Warning' begins with portentous vocal incantations announcing God's creation of the world, followed by its Fall at the hands of human beings: 'The foundations of the world are being broken'. Based around small intervals, long-held notes and a lack of repetition, the affectless vocal lines are unsettling. Axelrod's richly textured arrangement adds to the feeling of alienation by avoiding riffs and rhythmical *ostinati*, though Ernie Watts' tenor saxophone and Earl Palmer's propulsive rock drumming provide occasional but short-lived moments of release. The feeling of dislocation reflects the sense that 'the Earth is decaying around us'. The album's second part, 'The Sign', introduces more pastoral moments of flute and classical guitar, but continues the solemnity of 'The Warning': 'there will be nothing clean to drink for those who break the laws of nature'. The deep ecological sentiment is rearticulated in the album's closing moments, in 'the sacred words of Earth itself: it is lovely, it is lovely indeed'.

Overall, the didacticism of *Earth Rot* is enlivened by the complexly layered treatment of its text, and the music's refusal to provide the reassurances of rock music's verse-chorus structure. In contrast, Joni Mitchell used wit and humour to mediate environmental protest in an aesthetically satisfying way. She wrote 'Big Yellow Taxi' (1970) after a visit to the 'paved paradise' of Hawaii. The song protested against overdevelopment and the commodification of nature through its striking image of a museum which charges the public to see the last remaining trees. By making a witty parallel between the private loss of a lover and the public loss of America's green spaces ('You don't know what you've got 'til it's gone'), Mitchell ensured that her song moved beyond mere polemic, and would regularly be covered in every decade since its release.

The Beach Boys also viewed environmental apocalypse in terms of a fallen Pacific paradise. During the surf music era of the early sixties, the Pacific Ocean had been for the Beach Boys a space of innocent pleasure, celebrated in hedonistic songs such as 'Surfin' USA' (1963). After the Santa Barbara oil spill in January 1969, however, the ocean came to be seen as an endangered place. In 'Don't Go Near the Water,' the opening track of *Surf's Up* (1971), the characteristic Beach Boys' sound, as bright, cheery and optimistic as the Californian dream, sits uneasily with the ecological anxieties expressed in the lyrics. But the sweet tonal harmonies resolve the anxieties in a reassuring way, and the wordless vocalese prettifies the experience of environmental threat. As the key shifts upwards, a coda of mandolin, melodica and breathy 'oohs' ends the song literally on a high note. Seldom has environmental collapse sounded as sweet. The lyrics conclude with a generalised call to action and an appeal to liberal consensus: we should all 'help the water' by doing 'what we can and oughta'.

Two songs on Spirit's *Twelve Dreams of Dr. Sardonicus* (1971) also alluded to environmental collapse while providing therapeutic release from its anxieties. 'Prelude – Nothin' To Hide' sets rock's optimistic celebration of the potentiality of the individual ('the world is at your fingertips') against a recognition of ecological limits reminiscent of Malvina Reynolds in 1964: 'But see what you done to the rain and the sun'. The message is to 'wake up', before the chorus releases the tension in a boogie rhythm: the pleasures taken in music and a jug full of beer will solve any problem. The coda is a power chord vamp, decorated with fast slide-guitar glissandi, disorientating phase effects and punchy horns. Singer Randy California shouts 'Well alright', as the music becomes ecstatic escape.

'Nature's Way', in contrast, is a quietly meditative acoustic number which warns that the dying trees are nature's way of 'telling you something's wrong'. Dissonant harmonies and dramatic drum rolls mimic the impending environmental collapse, and the song ends with the sound of someone coughing as a car engine turns over, reducing environmental apocalypse to comic bathos.

Most topical rock songs of the era represented environmental apocalypse as a melodramatic struggle between 'Us' and 'Them', innocent David versus guilty Goliath. Spirit's 'Prelude – Nothin' To Hide' was an exception, in its use of the second person singular, as if directly addressing the listener: 'see what you done to the rain and the sun'. Such a rhetorical move was unusual; for the counterculture, error and fallibility tended to lie exclusively with the Squares. We have already seen Jim Morrison's cry of despair: 'What have they done to the Earth?' For Joni Mitchell, it was 'they' who 'paved paradise'. Captain Beefheart also blamed cruelty to humans and animals on an impersonal, unspecified 'they'. In 'Wild Life', he wants to escape into wilderness before 'they' take his life, his wildlife and his wife. He once explained that his pseudonym, given him by Frank Zappa in 1963, meant he had 'a beef in my heart against the world … The way they treat humanity and animals' (Trakin 1993: np).

By framing agency in this Manichean way, the counterculture tended to absolve itself of complicity in environmental degradation. Philosopher Kate Soper writes that the idea that 'we' are all to blame, that there is a 'general species accountability' for environmental damage, does not adequately distinguish between different human groups who may be more or less culpable (1995: 262). As writer Wendell Berry argues, the contemporary environmental situation is determined by both 'Them' and 'Us'. 'The world is being destroyed – no doubt about it – by the greed of the rich and powerful', he writes. 'It is also being destroyed by popular demand. There are not enough rich and powerful people to consume the whole world; for that, the rich and powerful need the help of countless ordinary people' (Hawken 1993:15). Yet sixties rock musicians tended to deny the commercial context in which their music was produced. As a result, questions of their complicity in consumer society, and consequently a serious, materialist critique of the environmental costs of that society, did not have to be faced.

## 7.4 Anti-pastoral as dystopian satire: Frank Zappa

Of all the rock artists of the 1960s, Frank Zappa alone was prepared to ask difficult questions about the presumed moral innocence of the counterculture's division of the world into 'Us' and 'Them'. As Kevin Courrier observes, Zappa 'gave audiences an image different from those beautiful groups who were trying to lull everyone back into the Garden of Eden' (2002: 86). His band the Mothers of Invention were instead 'the ugly reminder that railed against the optimistic wave of hippie idealism' (90). It is with Zappa's very different take on environmental issues that this chapter will end.

As a resident of Los Angeles, Zappa rejected the back-to-nature pastoralism of San Francisco bands such as the Jefferson Airplane or Quicksilver Messenger Service, satirising what he saw as the fake, 'more-rustic-than-thou' style of the San Francisco hippies (Charlesworth 1993: 34). For Zappa, only the anti-pastoral mode could do justice to what he saw as the atrocities of contemporary American society, including the pollution of the environment by the military-industrial state. An examination of his music in this period will show how he used modernist techniques to satirise the social processes at the heart of environmental degradation.

In Claude Debussy's 1894 orchestral piece, 'Prelude à l'Après-midi d'un Faune', the mythical creature lies half-asleep on a lazy summer afternoon, playing his flute and dreaming of nymphs. The Mothers of Invention's 'Prelude to the Afternoon of a Sexually Aroused Gas Mask', recorded live at the Royal Festival Hall in London with the BBC Orchestra in 1970, travesties Debussy's idyll, self-consciously using aural signifiers of racial and class division to expose the pastoral mode as an evasive white, middle-class construct. As Don Preston's organ part turns a quotation from Tchaikovsky's Sixth Symphony, 'The Pathetique' (1888), into lounge-bar Muzak, baritone saxophone player Motorhead Sherwood begins to 'snork' like a pig, a reminder of the animality of Debussy's faun. Although the sensuality of Debussy's piece was shocking for its day, the sound of its bodily lust was repressed in the genteel, high art mode of the pastoral.

The bathos and crude humour with which Zappa treated sexuality and the body was, in part, his means of questioning the transcendentalist mysticism adopted by the flower-power movement. His travesty of Debussy also questioned its racial politics. When Zappa calls out to Sherwood, 'Blow your harmonica son', the piece invokes the material reality of African-American poverty and hardship by quoting 'Trouble Everyday', his own blues song about the Watts riots. Zappa references the blues to point to the realities of racial oppression that the myth of pastoral contentment complacently ignores.

'Prelude to the Afternoon of a Sexually Aroused Gas Mask' also highlights the realities of pollution caused by the military-industrial complex. The title evokes Zappa's childhood memories of the gas masks his parents kept at the family home near Edgewood Arsenal, Maryland, in case of spillage from the tanks of mustard gas at the military base where his father worked as a meteorologist (Watson 1994: 19). Such images of despoliation recur in what Zappa calls his 'conceptual continuity'. He thought of the title 'Nine Types of Industrial Pollution' (1969) after driving through New Jersey. In 1993, he released 'Outrage at Valdez' for a Jacques Cousteau documentary about the Alaskan oil spill of 1989. In both instrumental pieces, modernist dissonance expresses the shock and ugliness of social and environmental damage.

Zappa's concern for the damaging effects of industrial pollution was unusual because it had none of the nostalgia for arcadian nature that characterised his contemporaries in the rock field. Indeed, his work dwelt on, and even relished, the ugliness of the natural world, and its potential hostility towards human beings. He thereby reminded his listeners that the desire to go back to pristine nature is a Romantic fantasy that masks the material facts of labour, power and class inequality that have been fundamental to human relationships with the natural world, and cannot be wished away. The brutality of nature is evident in Neon Park's cover art for *Weasels Ripped My Flesh* (1970), on which 'Prelude to the Afternoon of a Sexually Aroused Gas Mask' was released, which shows a man cutting himself shaving, using a weasel instead of an electric shaver. The confusion of animal and machine is typical of the conflation of the organic and inorganic in

Zappa's work, a destabilising of the opposition between nature and culture also evident in the image of the 'sexually aroused gas mask'.

For Zappa, human contact with nature is always mediated by the necessity of labour. 'Kids have a natural sense of mysticism, and a feeling of being connected to nature', he wrote in 1989. 'The natural world is very exciting when it's all brand-new.' He then added a typically sly sting in the Romantic tail: 'For example, kids have an appreciation for snow which is generally not shared by the guy who has to shovel a driveway. The older you get the more you take nature for granted (unless you're Euell Gibbons and you want to eat everything that's lying in the ground until you die from it)' (Zappa 1989: 255). For Zappa, then, the instrumentality of human relationships with nature is inevitable. Nature has always been a site of labour and its attendant social inequalities. The satire on Euell Gibbons, the 'Father of Modern Wild Foods', rejects the automatic association of nature and the 'natural' with good health in New Age culture. Zappa made a similar point in the Royal Festival Hall concert of October 1968, released on *Ahead of Their Time*, in a skit in which macrobiotic food turns Don Preston, dressed in a villain's cape and top hat, into a homicidal monster.

The ruination of nature epitomised for Zappa not a loss of Edenic innocence, but the repressiveness of the military-industrial complex and the consumer society it maintains. In a concert in Stockholm in 1967, he introduced his instrumental piece 'King-Kong' by saying that the gorilla was 'doing okay' in the jungle before the Americans came and 'made some money – by using the gorilla – then they killed him' (Courrier 2002: 159). Siding with this epitome of wild nature did not mean that Zappa fantasised about going back there. Instead, he rejected the naive anti-urbanism that proved so limiting to the environmental politics of the counterculture, and its subsequent development into radical environmentalism.

Unlike the San Francisco bands from which he took great pains to distance himself, Zappa admitted his complicity in consumer society, rather than feigned transcendence from it. Nor did his work succumb to the elitist dreams of escape into pastoral nature we saw earlier in this chapter in Paul Kantner's 'Hijack' and David Crosby's 'Wooden Ships'. Getting back to nature in such a politically evasive way evidently did not appeal to the resolutely urban Zappa, who told a story of how he gave up smoking during a chest cold: 'I walked into a hotel room and I could smell the rug, the disinfectant – as for smelling "the great outdoors", I don't live outdoors. Outdoors for me is walking from the car to the ticket desk at the airport' (Courrier 2002: 159).

Zappa's anti-naturalist stance may be cynical and irresponsible; if the biologist E. O. Wilson's 'biophilia' hypothesis is correct, and the emotional affinity that human beings have with the natural world is part of our evolutionary development, then he is in denial (Wilson 1984). Nevertheless, his work serves as a useful provocation to highlight some of the Romantic blindspots within the utopianism of the sixties counterculture, and as such anticipated the anti-naturalism of punk rock, which we will look at in Chapter 9.

Yet even Zappa's music shared the therapeutic function typical of rock, particularly in its self-consciously zany humour. One of his musical heroes, after all, was Spike Jones and his City Slickers. When sixties rock songs drew attention to serious issues, such as environmental degradation, they provided emotional and physical release at the same time. In doing so, they may be seen to fulfil in various degrees the role Ernst Bloch attributed to concrete utopias: hope for a better society in the future, based on a new, relatively benign relationship between human beings and the natural world, combined with a generalised critique of the social and political conditions that were putting that world in danger.

# Chapter Eight

# Country rock

The sixties rock culture took some of the dread off its fascination with technological futurism by simultaneously embracing old Americana. R. Cobb's cover for Jefferson Airplane's *After Bathing At Baxter's* (1967) depicted the band's eponymous 'All-American flying machine', a multi-coloured bi-plane shaped like an old San Francisco townhouse, rising above a squalid, monochrome cityscape of factory chimneys and wrecked cars. With its exuberant balloons and banner, the Jefferson airplane is an image of transcendence through convivial, homespun technology. By the late 1960s, such nostalgia for old Americana had become a new musical fashion: country rock. The Byrd's *Sweetheart of the Rodeo* (1968) left behind Roger McGuinn's science fiction modernism for an entire album of country stylings; on The Grateful Dead's *Workingman's Dead* (1970), Casey Jones' railroad train replaced the jet aircraft noise and spaceship imagery of the band's earlier work.

In his article 'The Topical Song Revolution and How it Fizzled Out', Irwin Silber, the Marxist editor of *Sing Out*! magazine, epitomised the discursive and experiential shift that was taking place. In 1965, he had opposed Bob Dylan's move into rock, but by September 1968, he had changed his mind, not only about Dylan, but about rock music as a whole. Showing the influence of Theodore Roszak's writings, which had appeared in the *New Republic* earlier that year, Silber belatedly celebrated rock as the music that most captured the aspirations of the youth counterculture. 'If ever an art reflected the rejection of a society's most sacred values, the rock music of the 1960s does', he wrote. 'This is the generation which has come to understand the nature of the crimes committed in the name of intellect, logic, technology.' Silber went on to reiterate the key ideas of the counterculture: 'We will reject wars and replace them with decibels, we will reject logic and replace it with mind-expanding drugs, we will reject intellect and replace it with the guts and the groin, we will reject technology and replace it with the natural and the animal' (Silber 1968: 21).

The Romantic opposition that Silber enunciated here between evil technology and benign nature was widespread in rock culture at this time. Album cover art depicted the reassuringly natural and traditional, epitomised by Linda Ronstadt's *Hand Sown ... Home Grown* (1969), in which the singer is pictured on the front cover in a white see-through dress in front of a tree, and on the back, seated on a simple yellow wooden chair in front of an ivy-clad, stone house. The title summed up the disavowal not only of the modern industrial technology that made the record itself, but also of its commodity status. Both were erased in the rock culture's turn to 'nature', even if such rural roots were often fabricated.

Historians of popular music have accounted for the return to traditional forms such as blues and country in the late sixties in two main ways: an internal, formalistic development within rock music, and a response to the wider social and political situation at the time. Musically, the fusion of rock with country and blues styles was a return to formal simplicity, in reaction to what its advocates saw as the self-indulgent excesses and over-elaborate studio manipulations of the new 'art' or 'progressive' rock. While Brian Wilson of The Beach Boys saw The Beatles' *Sgt. Pepper's Lonely Hearts Club Band* (1967) as an anxiety-inducing masterpiece, for others, notably Bob Dylan, it was overproduced and pretentious (Heylin 2000: 184). His *John Wesley Harding*, recorded in the autumn of 1967, led the counter trend among white rock artists towards simpler, more acoustic, folk-based forms.

The wider social developments behind this aesthetic shift are more difficult to assess. There are three main ways of accounting for the emergence of country rock in this period: as escapist retreat, as a statement of political moderation, or as a prefiguration of a new environmentally aware society. The following pages deal with each of these theories in turn.

## 8.1 Country rock and the return to 'roots'

In material terms, the turn to 'roots' reflected the newly acquired taste for country living amongst the nouveau riche rock stars. Many had moved into upstate New York and the Californian valleys, which were the most popular areas ironically because of their proximity to the city.

Dylan bought a house in the artists' colony of Byrdcliffe, near Woodstock in New York State in 1965 (Heylin 2000: 203). By 1967, both The Grateful Dead and Quicksilver Messenger Service had left San Francisco for ranches in Marin County. There was also a large movement of musicians into the hills and canyons around Los Angeles, particularly Laurel and Topanga Canyons. Neil Young, Stephen Stills, Chris Hillman, Barry McGuire and Linda Ronstadt all moved to the latter, where horses complemented automobiles as modes of transport. Music publisher Dan Bourgeoise commented: 'One of the reasons people slipped out to Topanga was that you could pretend you were in Kentucky or Tennessee or wherever you wanted to be' (Hoskyns 1996: 187). By the early seventies, Aspen, Colorado had become the new voguish retreat for bands such as the Eagles and Stephen Stills' Manassas.

The music reflected this move into the rural. The white blues band Canned Heat followed their success at the Monterey Festival in 1967 with 'On the Road Again' and 'Going Up the Country', both hits in 1968. The latter typified the desire to retreat from the city, leaving behind 'all this fussing and fighting' for a rural idyll where 'the water tastes like wine'. Opting out of the game was necessary for survival. John Fogerty of Creedence Clearwater Revival went as far as creating an escapist fantasy of being 'Born on the Bayou', and spending a carefree childhood 'runnin' through the backwood, bare', even though he was born in San Francisco and had never even been to Louisiana (Doggett 2000: 103).

Was this return to roots music a conservative retreat from contemporary society into the reassurances of older cultural forms? 1968 had seen an upsurge in political violence in the United States with the assassination of Robert Kennedy and the police riot involving anti-war protestors outside the Democratic Convention in Chicago, while Richard Nixon's election as President indicated a conservative backlash against left-liberalism. As the anarchist Yippies and militant groups such as the Weathermen hijacked radical political protest, the New Left fragmented. For the counterculture, the innocence and optimism of 1967's Summer of Love in San Francisco were over. Accordingly, for some commentators, the easy pastoralism celebrated in country rock was a retreat from political engagement into insularity, complacency and self-centredness. Barney Hoskyns cites 'Safe in My Garden', written by John Philips of The Mamas and the Papas in the summer of

1968: the world was 'on fire', but 'safe in my garden an ancient flower blooms'. In 'Our House', on *Déjà Vu* (1970), Graham Nash also sang praises to his new house, his cats and his easy life. Hoskyns writes that Crosby, Stills, Nash and Young's journey from Neil Young's 'Ohio', an angry protest against the killing of students at Kent State University by the National Guard in May 1970, to 'Our House', released as a single in September 1970, only a month after Young's song, 'seemed to sum up a general failure of nerve in the LA music scene' (Hoskyns 1996: 198). The general retreat into apathy and political quietism was epitomised by the words of the British blues-rock band Ten Years After in the fall of 1971: 'I'd love to change the world, but I don't know what to do, so I'll leave it up to you'.

For rock journalist Lester Bangs, this privileged retreat into the pastoral was led by James Taylor, the death of all things Dionysian in rock. Bangs considered his songs to be sappy, egocentric and anti-social, as his review of Taylor's album *One Man Dog* in February 1973 made clear: 'It's James at home, a one-man parade indeed, and as such a monolithic archetype for our time. He doesn't care about anything in particular except himself, the love he's found, his dog, and the lanes and pastures in his neighbourhood which he finds great contentment ambling through' (1987: 115). Bangs ends his review by giving Taylor some advice: 'come on outa the closet, James: stop trying to be the J. D. Salinger of the count-out culture, slouch on down and drool by the lamppost and the bar with the rest of the wetbacks. You'll be an even greater American and more of an inspiration than you are now.' Bangs defends the urban values of the rock culture – universality, hedonism and sociality – against Taylor's apparent withdrawal into the rural.

For sociologist Todd Gitlin, a former president of Students for a Democratic Society, the return to musical 'roots' at the end of the sixties also mirrored the growing political quietism and conservatism of the protest movement. Instead of the 'millennial pitch' of The Doors, Jefferson Airplane and The Rolling Stones, what had developed was 'a music to coax the breakneck imagination down to earth – even if that meant inventing an America in which one could sink roots, and a self capable of sinking with them there' (Gitlin 1987: 428). As Peter Doggett comments, 'Based securely within corporate America, self-confessed revolutionary musicians of the late sixties could idealise the pull of the land […] at the same time as they milked multi-national corporatism for all of the riches it could offer' (2000: 164).

The record industry sold the concept of 'the country' as a fashionable commodity for urban audio-tourists. In January 1969, Warner Brothers-Reprise's advertising executive Stan Cornyn had the idea of giving away a free bag of Topanga Canyon dirt to anyone who bought the first Neil Young album (Hoskyns 1996: 200). When the Flying Burrito Brothers' *The Gilded Palace of Sin* was released in March 1969, A&M Records sent journalists advanced copies in parcels stuffed with hay (Doggett 2000: 72). But although Crosby, Stills, Nash and Young described their sound in naturalistic terms as 'wooden music', for Langdon Winner, reviewing *Déjà Vu* for *Rolling Stone*, the album's faux-leatherette cover was an indicator of what was within. The album, he wrote, 'would like to convince you that it has roots deep in the American soil. But a closer inspection reveals that its tap root is firmly implanted in the urban commercial asphalt' (Doggett 2000: 53).

There is, however, an alternative, more positive interpretation of country rock than this, which was developed by some commentators writing both at the time and in retrospect. According to this view, it was not so much a retreat from politics into rural chic as a statement of necessary moderation in the increasingly polarised and violent society of late sixties America. The Nitty Gritty Dirt Band's 1971 triple album *Will the Circle Be Unbroken*, for example, was a collaboration between the new hippie country musicians and old-timers such as Earl Scruggs, Maybelle Carter, Merle Travis, Doc Watson and Roy Acuff. Reviving styles from the 1920s and '30s, the record was thus a gesture of reconciliation between the warring generations, the hippies and the squares: the cover quoted Chet Flippo of *Rolling Stone* and Jack Hurst of the *Nashville Tennessean* on the meeting of two worlds.

Contemporary reviewers and critics also wrote of Bob Dylan, The Band and The Byrds as voices of moderation amid the fractious politics of 1968. In his review of *John Wesley Harding* for *Crawdaddy* in April 1968, Jon Landau praised Dylan's new-found ability, as he put it, to 'sustain an identity of being a "moderate man"', which placed him 'much closer to approaching the tradition of the genuine folk-artists whom Dylan modelled his career after in the beginning' (Landau 1968: 120).

Like Dylan's return to acoustic music, The Band also set out deliberately to make music that was an alternative to the social and cultural developments around them. However, whereas Dylan, as we will see later in this chapter, was typically ambiguous, The Band, made up mostly of Canadians, unequivocally celebrated the rootedness of life in small town America and the lost values of the nineteenth-century frontier. The sepia-tinted cover photograph on their debut album *Music From Big Pink* (1968) showed the group in working men's clothes; the inner sleeve depicted them with their family and friends in front of an old wooden-framed house. 'Against a cult of youth', writes Greil Marcus, 'they felt for a continuity of generations; against the instant America of the sixties they looked for the traditions that made new things not only possible, but valuable; against a flight from roots they set a sense of place' (1997: 45). The Band's instrumentation of fiddle and mandolin harked back to 1920s string bands, and established a sense of continuity with the past. Moreover, the group's playing emphasised a collective sound that spoke of community rather than the competitive, egotistical individualism of rock.

This interpretation of country rock as a voice of political moderation should be considered in the context of a third interpretation of the music, as an expression of the newly emerging ecological awareness. 'Respect for the planet', writes Peter Doggett, 'encouraged a vague environmentalism; equally pervasive was the feeling that country, the sound of the soil, echoed this shift in perception' (2000: 139). People turned to the country in this period, David Downing argues, 'not merely as an escapist manoeuvre, but also towards new values that the country and its music might have to offer' (1976: 136). From this perspective, even James Taylor can be redeemed from his savaging by Lester Bangs. Doggett comments that 'royalties, rather than family tradition, might have earned Taylor his "homestead on the farm", but his early albums evoked a far closer identification with the land and its values than Ray Price or Eddy Arnold were offering at the start of the decade' (2000: 333).

We saw in the previous chapter that apocalyptic fears of environmental decline surfaced in many rock songs of the late sixties and early seventies, in The Doors, Captain Beefheart and Spirit, among many others. Country rock also reflected such anxieties. Two songs on the New Riders of the Purple Sage's eponymous first album (1971) connected environmental fears with the failure of the counterculture's desire to return to nature. 'Garden of Eden' is a quiet lament for an environmentally damaged world, punctuated by Jerry Garcia's plaintive cries on pedal steel guitar. The lyrics tell of smoke-filled air and blood spilled on the ground. The tree is not as green 'as it used to be'; the water not as 'cool' and 'clear' as it ought to be. The chorus reminds the listener that we live in a Garden of Eden: 'Don't know why we want to tear the whole thing down'. Unusually for topical environmental songs in this period, the song ends by acknowledging the need to overcome the distance between the hip counterculture and the rest of the American people: 'Hey, look at the people / Sometimes we forget that we're just people'.

The other song with an environmental theme, 'Last Lonely Eagle', directly faced the failed utopian hopes of the counterculture hinted at in 'Garden of Eden'. The people who live 'round the bend in the river' have 'forgotten their dreams' and 'cut off their hair'. Instead of the peacefulness of life in the mountains, there now remains only the 'gas powered flat land / Where most of the people just think that they're free'. But the last lonely eagle will never land, like the people's dreams of freedom.

'Last Lonely Eagle' mourned the decline of the back-to-the-land movement of the late sixties. At this time, some members of both the New Left and the counterculture had believed that rural communalism was a potential social and economic solution to the nation's growing environmental problems. Communes would create an alternative economy based upon subsistence agriculture and less wasteful, small-scale production. While some communards stayed in the cities, the 'back-to-the-land' movement accelerated after 1968. Musicians played an important role in promoting the rural commune movement. Lou Gottlieb of the folk group the Limeliters founded the Morning Star commune near San Francisco in 1966, while The Grateful Dead later supported a commune near Mendocino. The most ambitious attempt to establish an environmentally benign rural commune was, however, the Earth People's Park project, formed by a broad coalition of hippies, the Hog Farm commune, the Ecology Action Council, the Sierra Club and underground newspapers. David Crosby was among the rock stars to give financial support to the venture. The underground press asked readers each to send a dollar towards purchasing land for a permanent living space in either New Mexico or Colorado. A flier read: 'EARTH PEOPLE'S PARK is not a music festival / is work and love and generosity and devotion and play' (Anderson 1995: 226-7). Like most communes, however, the venture was short-lived, eventually defeated by a combination of local opposition, police repression and financial hardship.

More than any other songwriter, Jackson Browne faced up to the failure of the counterculture's back-to-nature utopianism. His songs captured the confusion and alienation of the period after 1968, as they vacillated between a continuing desire for radical social change, and an anxious realisation that, in David Downing's words, 'the old society is either too secure or too far gone for change to be a social possibility'. Browne explored contemporary moral dilemmas: 'to leave or to stay? To fight or to opt out? A social or a personal solution?' (Downing 1976: 142). 'Take It Easy', the opening track on *For Everyman* (1973), fantasises about a feeling of ease beyond this 'world of trouble', and recognises the difficulties of attaining it. The advice in the chorus is simple: to 'lighten up' and not even 'try to understand'. In this case, the typical rock message of *carpe diem* translates into a sexual encounter with a girl in a flatbed Ford in Winslow, Arizona. But the music itself undercuts the sense of ease encouraged in the chorus, as Russ Kunkel's double-time cymbals suggest nervous energy, and Clarence White's spiralling runs on the pedal steel guitar become increasingly convoluted.

The emphasis in 'Take It Easy' on personal redemption is questioned by 'Our Lady of the Well', into which it segues. This song explores more fully the notion of finding 'a place to make your stand' mentioned in the previous track, by dramatising the narrator's ambivalent feelings on visiting a rural commune in New Mexico. It opens with the sanctity of a simple action: drawing water in the open air becomes 'a dance we do in silence'. But the communards' attempt to work the land 'as they have always done' is doomed, because it is 'so far the other way my country's gone'. The 'cruel hand' of political reaction in Nixon's America threatens the simple values of working the land. Moreover, even though the narrator admires the commune, he must return to the city, where his 'heart remains among them'. As an artist, he has a duty to his public, and staying in the commune feels like a retreat.

The next track on the album, 'Colours of the Sun', evokes an idyllic pastoral scene beyond materialism and commerce. It is a paean to the synaesthesia of psychedelic experience: 'Voices in the air, sympathetic harmonies / Coming from the trees'. The Blakean vision of wonder is neither self-deception nor a dream, but an alternative to the reductive, patriarchal world of reason and quantification, in which 'Dying men draw numbers in the air / Dream to conquer little bits of time'. Browne's voice holds the final note for a few extra beats, enacting the sense of slow ease he has found in the commune. However, as in the previous track, visiting the commune again convinces him that his proper place is in the city. As the drums become ominous and heavy, he decides that he does not want his life to be restricted by a single plan. The communards are searching for 'the way', but the narrator is a fallen man, unable to find salvation through such means.

After these two songs directly addressing the rural commune movement, the rest of *For Everyman* refocuses on more personal issues: hopes and failures in love, and conflicting desires for separateness and intimacy. However, the title track, which closes the album, returns to the social orientation established at the start, and ties together its themes. People are preparing to escape an impending apocalypse, having 'heard their last warning'. But the narrator cannot retreat to the rural communes and 'give up the race' with a small group of friends, because such a life is not available for everybody. He refuses to abandon 'Everyman', the democratic collectivity, even though he is uncertain whether 'he'll show'. He is non-dogmatic, and does not claim to have an answer ('If you see somewhere to go I understand'), but doubts that self-reliance and rural retreat are solutions to the impending catastrophe.

Browne's *Late For The Sky* (1974) was also haunted by dreams of escape from an impending but unspecified environmental apocalypse, as the album explores complex feelings of regret, guilt, confusion, resilience and a yearning for transcendence. In 'The Road and the Sky', dark clouds will 'wash this planet clean just like the Bible says'. But the song is an up-tempo escape fantasy celebrating mobility for its own sake, as the narrator takes to the road in a stolen Chevrolet, 'rolling away from yesterday' and aiming to 'hot-wire reality'. Again, the message is to seize the moment rather than plan for an uncertain future. 'For a Dancer' filters similarly apocalyptic fears and confusion through the therapeutic, happy-sad mood of an old-time Appalachian fiddle tune: 'Don't let the uncertainty turn you around', Browne sings, 'Go on and make a joyful sound'.

In 'Before the Deluge', there is a sense of belatedness, because the unspecified catastrophe has already happened. The counterculture's utopian dream of going back to nature has failed. Environmental activism has proven futile, and those who were angry at the abuse of the Earth by 'the men who learned how to forge her beauty into power' find that they themselves have become victims of 'the magnitude of her fury in the final hour'. The rural communards who innocently sought 'their journey back to nature' cannot themselves escape the deluge. The chorus seeks out optimism from this pervasive sense of fatalism and loss, and finds a therapeutic value in music itself as a means of keeping the spirits high. The song ends with a vague hope of mystical deliverance, that 'creation' will 'reveal its secrets by and by'.

Browne's songs explored the pull between egocentrism and social awareness, and the question of how to preserve hope for progressive change in a damaged world. In the late 1970s, as we will see in the next chapter, he became an anti-nuclear activist. For Browne, as for Don Henley, drummer and songwriter with The Eagles, whose work we will also explore further in the next chapter, the values of country rock informed their emerging environmentalist sensibility. Speaking in the 1980s, Henley elaborated on the link between country rock and environmentalism. 'That whole movement', he said, 'was connected to environmentalism, because it was a music that had grown in part out of country music. It was very much connected to the earth, and everybody was wearing earthy clothes and celebrating the outdoors. It was a very natural time. And it all made sense with the music. Then in the late seventies, there was a backlash against that. The country, the natural sound that is connected with nature, has gone out of music pretty much' (Doggett 2000: 165). Henley was referring to the rise of disco and punk in the late 1970s, which, as we will see in the next chapter, largely rejected the rural values associated with country rock.

If Henley displays a conviction about the moral value of rural life, other country rock performers were less sure. Journalist Robert Christgau wrote in November 1968 that country music was 'the gimmick of the year', and accounted for its popularity in the appearance of 'traditional honesty' it gave in an otherwise fallen world of urban commerce. He also observed that irony played an important part in the new country rock performances, noting that the 'line between high camp and real country music is often nonexistent' (Doggett 2000: 86). Irony allowed rock musicians to play country music and remain hip. They could draw back from conviction, and leave it to their audience to decide how their material should be received. Irony also suggested that the pastoral vision within country rock was insecure. As we will see in the final section of this chapter, this sense of irony pervaded the country-influenced music of Bob Dylan in this period.

## 8.2 Bob Dylan's nervous pastoral

The 'message' of Dylan's late sixties work, according to Gregg Campbell, was that 'if man cannot achieve salvation through the ever-escalating domination of self, fellows and nature, then the important thing, the essential thing, the only truly human and noble course is to attempt to live a flawed life in communion and harmony with one's fellows and one's natural environment' (Campbell 1998: 105). Yet Dylan's country rock songs were about the difficulties of achieving this harmony with the natural world. 'Strap yourself to a tree with roots', he sang in 'You Ain't Goin' Nowhere', officially released on *The Basement Tapes* in 1975: for a typically sly and ironic Dylan, going back to nature is more an act of wilful, self-enforced immobility than of freedom. The concerted effort required to return to static roots only indicates the distance between human beings and the natural world. Indeed, for Dylan, the country was never an unequivocal escape from the urban corruptions he explored in mid-sixties songs such as 'Desolation Row' (1965) and 'Visions of Johanna' (1966).

Dylan's turn to country music began towards the end of *John Wesley Harding*, an album on which he explored his familiar humanist concerns with individual guilt, responsibility and redemption, this time through the outlaw imagery of the American West. Towards the end of side two, the people give their advice to Dylan's alter-ego, the Wicked Messenger: 'if ye cannot bring good news, then don't bring any'. The next track abruptly brightens the mood: 'Down Along the Cove' is an upbeat piano boogie-woogie that announces the moment that Dylan, in the words of Wilfred Mellers, 'relinquishes his confrontation with doom' (1984: 158). The 'good news' Dylan brings is redemption through love, both sexual and spiritual, and is marked stylistically by a return to the simplicity of 1950s rock 'n' roll and country music.

The turn to the pastoral in the closing tracks of *John Wesley Harding* indicated a withdrawal from the dread encountered in earlier songs such as 'All Along the Watchtower', which exuded a vague sense of apocalyptic threat, as two mysterious riders approach a medieval watchtower in a howling wind. But with Dylan, irony always undermines the possibility of pastoral retreat. In 'Down Along the Cove', as Mellers observes, Dylan sings 'what for him is an "artless" lyric' in a 'rarefied, airy timbre, as though he is looking wonderingly at rather than entering into the experience' (1984: 158). The cheeriness of the melody is undercut by blue notes and uncertain rhythms, which prevent the song from being an unequivocal celebration of both erotic pleasure and the natural world.

The album's closing track, 'I'll Be Your Baby Tonight', is equally ambivalent. The lovers try to forget what cannot be changed in the outside world, represented by images of nature: the mockingbird and the 'big, fat moon'. But this acceptance of natural limits can be interpreted as either liberating or evasive, just as the song's nursery rhyme quality can suggest either spiritual rebirth or the bathos of infantile regression, the child-like or the childish. Despite the narrator's claim that all fears have been banished, an undercurrent of tension remains in Dylan's strained vocal delivery, which adds an element of pain and vulnerability to the optimistically rising melody. The closing harmonica solo sobs and wails with despair. Dylan's performance of the pastoral thus includes an uneasy recognition of decay and death, his version of *Et in arcadia ego*, in keeping, as we saw in Chapter 6, with his understanding of the sense of irrational 'mystery' at the heart of Anglo-American folk music (Mellers 1984: 159). In an interview in March 1966, Dylan had said he looked forward to 'Salvation. Just plain salvation' (Heylin 2000: 243). *John Wesley Harding* ends with a self-deluding attempt to find salvation in pastoral nature.

After a delay of nearly eighteen months, Dylan's next officially released album, *Nashville Skyline* (1969), showed him further pursuing what Mellers calls his 'path of relaxation', as he fully embraced the country stylings introduced at the end of *John Wesley Harding*, and at the same time excluded the songs of inner struggle that gave emotional and intellectual depth to the rest of that album (Mellers 1984: 161). Nevertheless, for Dylan, relaxation into the pastoral was again not a simple matter. Instead, *Nashville Skyline* contained disruptive nuances, in lyrics, music and vocal performance, that again undermined the pastoral idyll.

By 1968, Dylan was finding life in Woodstock difficult, because of what he later called the 'wave of insanity breakin' loose' around his house, with fans pestering him at all times of the day and night. This was, he added, 'just about the time of that Woodstock Festival, which was the sum total of all this bullshit' (Heylin 2000: 307). Dylan's sense of ambivalence towards rural life, and its embrace by the counter-culture, may be sensed in *Nashville Skyline* itself. The cover photograph shows him grinning and doffing his hat, playing the part of the Nashville family entertainer. The back jacket photograph shows the unremarkable, and decidedly urban skyline of Nashville, appropriate for a country album in which there is little attempt at rural authenticity or even rustic imagery.

The album itself further suggests impersonation or masquerade, as Dylan toys with his star persona. His use of language is faux-naif in its use of commonplace and banality. The use of cliché and of obvious Tin Pan Alley rhymes ('arms'/'charms') is playful and knowing. Love makes the world go around, Dylan sings in 'I Threw It All Away', before adding: 'it can't be denied'. Truism is presented as folk wisdom. In 'Lay, Lady, Lay', the directness of the language reflects the virtues of honesty and dignity associated with rurality and manual labour: 'His clothes are dirty but his hands are clean / And you're the best thing that he's ever seen'.

'Country Pie' is the album's most expansive and unambiguous paean to the joys of country living. The song lightly celebrates opting out of competitive, urban society into a country life in which voluntary simplicity is natural abundance: 'I don't need much and that ain't no lie / Ain't runnin' any race'. The country is a place of honest vulgarity and good humour, and the lyrics relish colourful rural nouns: goose, peach, raspberry, strawberry, lemon, lime, blueberry, apple, cherry, pumpkin and plum. But the album ends with 'Tonight I'll Be Staying Here with You', a close rewriting of *John Wesley Harding*'s finale 'I'll Be Your Baby Tonight', which reiterates that song's desire for escape from the public world. But the feeling of stasis and repetition only points to the limitations of rurality in Dylan's work. Once in the country, it seems, there is nowhere else for him to go, except back to the city.

*Nashville Skyline* divided fans and critics alike. For some in the New Left, it signalled political retreat. Robert Christgau, reviewing the record for *Village Voice*, was uneasy about Dylan's association with 'enthusiastic Nixon supporter' Johnny Cash, and had reservations about his embrace of a music which was, he said, 'naturally conservative [...] intensely chauvinistic, racist, majority-oriented and anti-aristocratic in the worst as well as the best sense' (Doggett 2000: 15-6). In contrast, Charles Reich wrote approvingly in *The Greening of America* of the 'serene, but achieved, innocence' of Dylan's record, while some members of the New Left even managed to find political leadership in Dylan's turn to country music (Reich 1971: 187). In his article 'Bob Dylan Meets the Revolution', published in October 1969, Tom

Smucker claimed that *Nashville Skyline* demonstrated the singer's new found solidarity with the rural working class, and therefore made an important link to the missing constituency in the New Left's Revolution. Smucker argued, rightly, that one of the main reasons for the failure of the rock culture's dream of an alternative community was that its elitist emphasis on youth culture had limited its political potential. Dylan's *Blonde on Blonde* (1966) typified for Smucker this 'detached, exotic world'. Compared to the self-righteousness and individualism of Dylan's mid-sixties work, then, *Nashville Skyline* showed a new sympathy for the working man (Smucker 1969: 181). If Dylan was carefully trying to undermine his leadership role by embracing country music, a style often reviled by left-liberals, he had clearly failed.

*Self-Portrait* (1970) showed even more ambivalence towards the pastoral tradition in country music. The inside cover shows Dylan looking up at the sky in sunlit woods. But the pastoral mode is again unstable. 'Belle Isle' uses stilted, medieval language ('For me there is no other damsel / Than my blooming bright star of Belle Isle') to distance the listener from the impossibly Edenic pastoral landscape depicted in the song. The western-influenced songs can similarly be taken as ironic. The album opens with 'All the Tired Horses', whose arrangement for strings and female vocals evokes the music of Hollywood westerns. But the repetition of the single line ('All the tired horses in the sun / How'm I supposed to get any ridin' done?') is disconcerting, suggesting stasis rather than the mobility and freedom usually associated with the American West. Added to the listener's vague feeling of unease is the non-appearance of the singer whose self-portrait this is supposed to be. The other western tracks on *Self-Portrait* similarly present an idealised pastoral threatened by bathos or irony. 'Wigwam' celebrates the vanishing American Indian in what Mellers describes as a 'wordless vocalise-pentatonic with microtonal arabesques', but, as he also states, the performance 'walks a tightrope between tragedy and farce' (1984: 165).

Hounded by his 'fans', Dylan and his family moved back to New York City in early 1970, only to find, as he put it, that the 'Woodstock Nation had overtaken MacDougal Street also' (Heylin 2000; 312). He recorded *New Morning* in the city between March and June of that year. The album refuses to be the unequivocal new start suggested by the title; again, there are hints of unease in the way the rural imagery is handled musically. 'Time Passes Slowly' is a country waltz in which anxieties surface because of its uncertain tonality and irregular rhythm. In 'Sign on the Window', although the narrator has lost his lover to another man, he still dreams of building a 'cabin in Utah', with a 'couple of kids who call me pa'. But his repetition of 'that must be what it's all about' suggests he is trying to convince himself. The tentative rhythm adds further doubt. The other tracks with a country theme are equally unconvincing. 'Winterlude' is another country waltz delivered in an exaggeratedly crooning voice, while the narrator of 'The Man In Me' discredits his country dream with a banal 'la-la-la' refrain. The rest of the album turns away from the country pastoral, in an unsuccessful quest for religious salvation.

The pastoral scene Dylan evoked in *New Morning* was too nervous to be truly escapist. Indeed, his encounter with country rock as a whole added an element of emotional complexity and ambiguity to a style that was often content to derive simple pleasures from the reassuring simulacra of nature it created. Other rock musicians, even more grounded in the city than the peripatetic Dylan, went further into the mode of anti-pastoral dystopia. Randy Newman's second album *12 Songs* (1970) included 'Let's Burn the Cornfield' and 'My Old Kentucky Home', both ironic demolitions of rural retreat. The former is an erotic ode to rural arson, born out of boredom and frustration. 'You hide behind the oak tree', the singer invites his lover, 'and I'll make love to you while it's burning'. Newman's travesty of 'My Old Kentucky Home' is a blackly comic catalogue of domestic violence, drunkenness and cruel violence to the natural world ('Shootin'' at the birds on the telephone line').

For Frank Zappa, country music was also synonymous with political reaction and ignorance. In 'Lonesome Cowboy Burt', Burtram Redneck is a redneck womaniser who loves beer and hates Communists. The song satirises cowboy mythology in a parody of a country waltz, complete with Jimmie Rodgers' yodel and comic horns. 'It Might Just be a One-Shot Deal' (1972) is a blues that slyly features 'Sneaky' Pete Kleinow, guest slide guitar player on many country rock records, in a strange, cautionary tale about a 'froggie' who dumps a large pile of sand onto a suburban rug. The sing-song chorus gives ambivalent advice on how to react when a 'forest grows up' out of the sand: 'you can be scared when it gets too real / But you should be diggin' it while it's happening'. Kleinow's pretty slide guitar interlude is brutally interrupted by an atonal, pulse-less passage which mimes the comic revenge of a violent, chaotic nature on bourgeois civilisation.

The dystopian anti-pastorals of Randy Newman and Frank Zappa serve as necessary correctives to the simple pastorals of Graham Nash and James Taylor. As David Downing argues about country rock in general, the decision 'to turn one's back on civilisation was not one to be taken lightly. Country rock, on the other hand, could be taken lightly as some sort of substitute for the real thing. If one couldn't actually go, then at least one could enjoy the yearning to go, the promise of blue skies and whispering pines and home-grown dope; above all the promise of somewhere else, an open frontier in the midst of all these doors closing' (1976: 137). In the more complex pastorals of Bob Dylan, Jackson Browne and the New Riders of the Purple Sage, however, the desire to live in harmony with the natural world came up against political and psychological realities which acted as a counter-force to such simplistic utopian hopes.

# Chapter Nine

# Post-1960s rock, R'n'B and hip hop

Rock music in the 1970s largely reflected the political quietism that replaced the protests of the sixties. However, some musicians continued to adopt political causes, including those in defence of the environment. By the early 1970s, environmentalism was one of a diverse range of political causes that had emerged from the fragmentation of the New Left and counterculture at the end of the sixties, as activists realigned their focus beyond the Civil Rights movement and opposition to the Vietnam War. In this period, pollution, urban welfare, the politics of food and hunger, nuclear power, wilderness conservation and the protection of endangered species became important campaigning issues. As we will see in this chapter, American musicians, including many who came to prominence in the sixties, participated in, and wrote songs for and about, these campaigns, and more generally contributed to the ongoing exploration of environmental issues in popular music.

## 9.1 Environmental protest in 1970s African-American R'n'B

By the late 1960s, African-American rhythm-and-blues or soul artists began to respond to their audience's desire for more overtly political lyrics. During the Civil Rights era, black audiences had created political significance out of otherwise apolitical songs such as Martha Reeves and the Vandellas' 'Dancing in the Street' and Aretha Franklin's 'Respect'. However, as Brian Ward writes, the 'demands of the black audience for more explicit racial and social commentary in soul lyrics coincided with the industry's recognition that not only was this a necessity in order to maintain credibility and sales among black consumers whose collective annual income topped $1,000 millions in the early 1970s, but that such songs would not automatically alienate the white market' (1998: 361).

R'n'B songs both celebrated black pride and culture and criticised poverty, inequality and discrimination. By the early seventies, such critiques began to include an environmentalist dimension, particularly in response to the social and ecological collapse of the inner city. Marvin Gaye's *What's Going On* (1971) extended opposition to the Vietnam War into a protest against environmental degradation. Gaye's move into social and ecological concern was motivated by the return of his brother Frankie from active service in Vietnam in 1967. The album, he said, was the product of three years spent reflecting on 'life and upon America especially – its injustices, its evil and its good'.

'Inner City Blues (Make Me Wanna Holler)' indicted economic inequality. The money wasted on space rockets could have been better spent on the 'have nots'. The tax system is also unjust: 'Money, we make it / 'Fore we see it you take it'. The simple, brief phrases of the lyric combine with sweet double-tracked vocal harmonies, layered polyrhythms and a catchy bass riff in an elegy that is both mournful and uplifting.

The album's main statement of environmental concern was 'Mercy, Mercy Me (The Ecology)'. The song is a threnody for the poisoning of the oceans and the land, in which the warmth of Gaye's voice against the floating, jazzy groove elevates the commonplaces of the chorus ('things ain't what they used to be') into therapeutic release. In an interview, Gaye elaborated on the Judaeo-Christian values which frame the record's social concerns. 'Find God: we've got to find the Lord', he said. 'Allow Him to influence us. I mean what other weapons have we to fight the forces of hatred and evil. And check out the Ten Commandments too. You can't go too far wrong if you live them, dig it' (Davis 1991: 117). In Gaye's hands, soul music thus developed from, but stayed close to, its stylistic origins in gospel, and could thereby offer a healing solution, both spiritual and physical, to loss and despair.

In 1971, Stevie Wonder set up his own production company which leased his records to Motown for release, and used this new artistic and economic freedom to explore social and political concerns in his music. 'Living for the City' uses a familiar narrative of black migration from the country to the big city to indict the exploitation of black people in the modern city. Set to an insistent beat, the verses tell of a boy born in 'hard time Mississippi', raised by loving parents who work long hours for little money. Unlike in Marvin Gaye's 'Inner City Blues', Wonder's lyrics explicitly put racial prejudice at the centre of economic oppression. Unable to get a job because 'they don't use coloured people', the boy walks the streets of New York City looking for work. Here Wonder's concern expands to include an apocalyptic view of the urban environment: he is 'almost dead from breathing in air pollution'. When the innocent country boy falls victim to the drug dealer, a sound effects sequence depicts a prison door clanking shut to the brutalising, racist insults of the jailer. In the final verse, the narrator draws the moral of the story, which again places the plight of the young black male within a scenario of environmental apocalypse: 'if we don't change, the world will soon be over'. As Brian Ward observes, there is 'no redeeming sense of a vibrant black culture, community or fellowship here; just the sordid and highly exploitative urban jungle which racism and poverty had created'. Nevertheless, as he notes of both Marvin Gaye and Stevie Wonder, 'either God or a more generalised, beatific "love" frequently served as an antidote to dissolution and despair' (1998: 366-7).

In 1979, Wonder released *Journey Through The Secret Life of Plants*, the album on which his ecological concerns came fully to the fore.The album is based on his soundtrack to the film version of Peter Tompkins and Christopher Bird's book, *The Secret Life of Plants*, which advocates both scientific and parascientific theories of plant behaviour. Mystical notions of an animistic world, the authors argue, may reveal insights into the workings of plants denied to the material sciences. 'The attraction of the seer's supersensible world, or worlds within worlds', they conclude, 'is too great to forgo, and the stakes too high, for they may include survival for the planet' (Tompkins and Bird: 1973: 318). Wonder's songs and instrumentals pick up on the New Age elements of the book, as he creates a hybrid of New Age electronica, world music and R'n'B'. For an album whose lyrics explore the organic imagery of plants and flowers, *Journey Through The Secret Life of Plants* marked the first use of a digital sampling synthesizer, the Computer Music Melodian, and was one of the first digital recordings. For Wonder, though, this combination of biophilia and technophilia was entirely unproblematic, as it has proved for many of the musicians featured in this book. Popular music becomes an exemplary model for the symbolic reconciliation of technology and nature.

Wonder's African-American electronic pastoral imagines plants and trees within a scientific and ecological context, and celebrates the symbolic role played by flowers in romantic love. The music invokes a 'One World' sense of cosmopolitanism and global unity that includes

India ('Voyage to India'), Japan ('Ai No, Sono') and Africa ('Kesse Ye Lolo De Ye'), signified by the use of sitar, koto and kora respectively. Polyglot children's voices reinforce the Romantic vision of natural innocence; cultural and biodiversity are shown to be part of the same ecological sensibility. At times mawkish and overblown, the musicality and inventiveness of Wonder's melodic and rhythmic sense, and the insight of some of the lyrics, nevertheless provides the album with aesthetic interest.

The opening instrumentals, 'Earth's Creation' and 'The First Garden', establish a Judaeo-Christian notion of Nature as a benign arcadia. 'Same Old Story' introduces the Fall from this state of perfection. By revealing that plants think, feel and move like human beings, Science itself has been taken 'by surprise'. The song indicts 'non-believers' who 'find what's real too hard to believe in', and follows the source book in celebrating the contribution to agricultural science of George Washington Carver, born into a slave family, who discovered that 'if farm land was to be rich / We must plant crops to replenish it' (Tompkins and Bird 1973: 125-31).

'Race Babbling' brings a sense of ecological endangerment to the album. Moving into harder disco-funk stylings, Wonder uses electronically distorted vocals and tonally oscillating synthesizers to give a queasy, apocalyptic feeling to his lyrics' indictment of 'Man's production / Life's corruption'. What will 'save you people' is the insight that 'life's connected'. So Nature teaches humility to hubristic Man, by confounding our normal, commonsense assumptions about scale: 'A seed's a star's a seed'. The album's closing sentiment picks up on the recurrent biocentric theme that, as the title track puts it, 'what we see as insignificant / Provides the purest air we breathe'.

Like Stevie Wonder and Marvin Gaye, Gil Scott-Heron wrote songs in the 1970s that related environmental issues specifically to African-American concerns, while at the same time appealing to universal humanist values. In particular, he wrote two eloquent songs against nuclear power. 'South Carolina (Barnwell)' (1975) indicted the environmental racism implicit in the government's decision to build a nuclear reprocessing plant in South Carolina, which would leave a 'great big time bomb ticking' in a predominantly African-American area of the country. The chorus called for a resurgence of the 'protests' and 'rage' for those 'dying in the jungles of Vietnam'. Scott-Heron's authoritative bass voice is complemented by choppy jazz-funk rhythms and angry saxophone interjections, as his vocal decorations enliven the didacticism of the lyrics.

'We Almost Lost Detroit' (1977), in contrast, is a slow, melancholy blues punctuated by synthesizer wails and discordant horns. The title referred to the accident at the Enrico Fermi-1 fast breeder reactor south of Detroit in 1966, in which there was a partial meltdown of fuel in the reactor core. Unofficial reports said that the authorities were preparing for an emergency evacuation of Detroit (Denselow 1989: 176). Scott-Heron asks why no-one has 'stopped to think about the babies / Or how they would survive?'

Gil Scott-Heron performed 'We Almost Lost Detroit' at one of the No Nukes concerts at Madison Square Gardens in New York City organised by Musicians United For Safe Energy (MUSE), the group formed in September 1979 after the nuclear accident at Three Mile Island in the March of that year. Jackson Browne, Crosby, Stills and Nash, James Taylor, Sweet Honey in the Rock, Tom Petty and Bruce Springsteen were among other musicians who took part. The campaign against nuclear power, of which MUSE was a part, largely succeeded in its objectives, at least in the short term: for over twenty years after 1978, plans for every new nuclear power plant were either cancelled completely or indefinitely postponed (Denselow 1989: 180). However, the issue of nuclear power remains a contentious one in environmental politics, with prominent British scientists James Lovelock and David King advocating its use as a better option than fossil fuels and as yet unreliable 'renewables' (Lovelock 2004). For activist musicians such as Bonnie Raitt, Jackson Browne and Graham Nash, then, the campaign is on-going: in 2007, they continued the anti-nuclear activism they first began in the 1970s by helping to organise NukeFree.org (Raitt 2009).

The No Nukes campaign demonstrates the wider involvement of American popular musicians in campaigns around environmental justice. For musicians such as Jackson Browne and Country Joe McDonald, environmentalism is a logical extension of the countercultural politics they developed in the sixties. It is to the development of such activist concerns in the older generation of sixties rock musicians, as they continued to make music in the 1970s and beyond, that we will now turn.

## 9.2 Rock music and environmental protest since 1970

As American rock musicians of the sixties generation moved into the seventies, wildlife conservation became a logical area for their activist concern. The Save the Whale campaigns of the 1970s attracted the support of Captain Beefheart, David Crosby and Graham Nash, and Joe McDonald, amongst others (Regenstein 1975: 59-73). On his 1973 British tour, Beefheart leafleted audiences in support of the recently formed Project Jonah. Farley Mowat's *A Whale for the Killing* (1973) was one of the few books he admitted to have read. Whales, as we saw in Chapter 7, were vital to Beefheart's creativity. One of his descriptions of whale song recalls Marshall McLuhan's notion of 'all-at-onceness'. 'Bizarre Records is putting out an album of whales singing,' he told a journalist, 'and, oh man, there's no way you can listen to it: just feel it, no special designated area, no point of interest. There is no point – have you ever seen water run square out of a round tap?' (Frame 1969: 3). For Beefheart, whales are superior musicians to human beings because their art is a product of intuition rather than reason, of nature rather than human artifice. 'I've heard their music', he said, 'and it's past trigonometry, calculus, past polygraphs and beyond that. They're smart, and it's frightening that we're killing them' (McGrath 1973: 2).

By the 1980s, Beefheart had become increasingly vocal about animal rights issues in general. His biocentric sensibility led him to dedicate *Shiny Beast* (*Bat Chain Puller*) (1980), in a typically non-dogmatic gesture, to 'all conservation and wildlife preservation organisations everywhere'. 'Apes-Ma', the poem that closes the album, is about an ape in a zoo who is too old to try breaking out of her cage like she used to, and is now reduced to 'going to the bathroom' too much. Beefheart protests against the mistreatment of animals in captivity through a subtle characterisation of the ape as a sentient individual, and consequently a figure of empathy and pathos.

Beefheart also made a public stand against vivisection and other forms of cruelty to animals. When he learned of Alice Cooper's alleged cruelty to animals on stage, he angrily warned him, via the music press, 'that if they met again he'd spank his ass' (Barnes 2000: 126). He said of whales: 'I definitely think that those things are very intelligent. They are not going to find out how intelligent they are by cutting into their brains, and all that bullshit that the scientists do under the auspices of inserting a ruler into their way of life' (Maltman 1972: 1-2). For

Beefheart, wild animals stood for a liberated, natural sexuality; conversely, human cruelty to animals was a symptom of sexual envy. He mocked actress Zsa-Zsa Gabor's desire for a tiger-skin coat: 'She must have a pussy problem. She's intimidated by the fact that the cat is so natural. It doesn't train itself. I feel people who wear cat skins have never really gotten away from mama' (Chorush 1971: 3). Beefheart's reference to dependency on the mother figure recalls his description of his music as an attempt to get away from the 'mama heartbeat', which we noted in Chapter 7. His music itself was a Romantic attempt to reconnect people with the natural energies which he believed were embodied by animals. At the end of 'Grow Fins', on *The Spotlight Kid* (1972), the comic narrator wants to escape domestic routine by reversing evolution and eloping with a mermaid: to 'grow fins' and 'get back in the water again'. For Beefheart, the only way to reconnect with primal nature is to regress to this pre-human state.

David Crosby and Graham Nash were another act from the sixties counterculture who extended their love for the natural world into activist concern in the 1970s. In his autobiography, Crosby tells of how learning to sail introduced him to the healing powers of nature. His schooner The Mayan, which he bought in 1967, stood 'for the good things in my life: health, sanity, and freedom – all the positive values'. Sailing taught him discipline and humility, qualities not usually typical of the rock star life he lived on land (Crosby and Gottlieb 1989: 128). Crosby had celebrated the ocean in song in 'The Lee Shore', on *Four-Way Street* (1971), which evoked the ocean as a pastoral sanctuary of infinite pleasure for the hedonistic self to expand into. All the way to Venezuela, he sings sweetly, are 'a hundred thousand islands / Flung like jewels upon the sea'. As the narrator relaxes on the deck in the sunset, he smells the dinner being cooked for him by the women below. If this is a touristic and patriarchal view of nature, by the mid-1970s Crosby's love of the ocean had led him to the anti-whaling campaigns that also attracted Captain Beefheart. Crosby and Nash's 'Wind on the Water' tour in 1975 raised money for Jacques Cousteau's organisation. In return, Cousteau gave the band some underwater footage for a promotional film (Denselow 1989: 271). The album, released in October 1975, included 'To The Last Whale', in which baroque vocal harmonies give way to rippling keyboard and harp arpeggios. This pastoral calm masks a scene of brutality and waste: the whales washed up on the shoreline have been killed for pet food and lipstick.

Disconcerting time changes make this elegy for wild nature nervous and unsettling.

Country Joe McDonald also refocused his countercultural politics towards environmental causes. 'Save The Whales!' (1976) was a jaunty sea shanty which ironically quotes the chorus from 'What Shall We Do With The Drunken Sailor?' to tell a history of his grandfather's whaling expeditions, before pointing the finger of blame at the contemporary Russians and Japanese, who scour the oceans 'Looking for ivory and perfume and plastic toys and pet food'.

As the anti-whaling campaigns continued into the 1980s, the plight of the whale also marked the development of Lou Reed from urban flâneur to satirical moralist. In 'The Last Great American Whale' (1989), Reed's characteristically affectless voice edges into anger as he narrates the story of the chief of an Indian tribe in the Carolinas who kills the racist son of the local mayor, and is rescued from Death Row by a great whale, summoned out of the sea by his 'tribal brothers'. The whale is shot dead by 'some local yokel member of the NRA', who mistook it for the Indian chief. The moral of this sardonic ballad of violence and stupidity is that Americans 'don't care for much of anything / land and water the least'. They will 'shit in a river, dump battery acid in a stream'.

In 1982, the International Whaling Commission adopted a moratorium on commercial whaling, an agreement which remains under constant pressure from pro-whaling nations such as Japan and Norway (Stoett 1995: 134). As the fate of the whale faded somewhat from public attention in the mid-1980s, the destruction of the rainforests became the new popular cultural emblem for an endangered planet, even inspiring The Grateful Dead to make their first explicitly political public stand. 'Although (Jerry) Garcia had spent years telling people that he wasn't a leader for any movement', observes Sandy Troy, 'the importance of this issue for him outweighed his career-long philosophy' (Troy 1994: 230). In 1984, the band had founded the Rex Foundation, a non-profit charitable organisation whose aims include 'supporting environmental, social, and cultural endeavours in order to secure a healthy environment, assist those in need, and ensure the cultural survival of indigenous peoples around the world' (Denselow 1989: 202). In 1987, Garcia licensed the use of his name for Ben & Jerry's 'Cherry Garcia' Ice Cream, and included royalty payments both to himself and to the Rex Foundation. Again displaying his continuing allegiance to hip capitalism as a means of social renewal, he also did a radio advertisement for Levi 501 jeans, on which he sang that 'A good pair of Levi's are bound to set me free' (Troy 1994: 224).

In September 1988, The Grateful Dead held a press conference on the state of the rainforests at the United Nations in New York. Next to a specially designed rainforest painting by Robert Rauschenberg, Garcia announced that, as 'a citizen of the Earth' he objected to the destruction of the rainforests. His speech acknowledged the 'multitude of political, social, and ecological factors' that was affecting the rainforests, concluding: 'It's going to take a lot of people – here in America, in Japan, in western Europe – working to change things at the consumer level, with official government policies, and corporate and international banking policies that finance rainforest destruction' (Troy 1994: 230). The band played a benefit concert at Madison Square Gardens for Rainforest Action Network, Cultural Survival and Greenpeace. The following year, Garcia, Bob Weir and Mickey Hart went to Congress to give testimony at the congressional Human Rights Caucus on the impact of the timber and meat industries on rainforest depletion.

The Grateful Dead's music of this period reflected these environmentalist concerns, particularly as interpreted in the video *So Far* (1987), directed by Jerry Garcia and animator Len Dell'Amico. The video for 'Rhythm Devils', a live percussion piece by Mickey Hart and Bill Kreutzmann, recalls Godfrey Reggio's *Koyaanisqatsi* (1983), by contrasting images of 'primitive' spirituality (Stonehenge, the Pyramids and desert nomads) with an environmentally destructive modern industrial society (skyscrapers, smokestacks, gridlock on a freeway and office workers in the city). The polyrhythmic energy of Hart's percussion suggests that closeness to the organic rhythms of nature is a key to social and environment progress. 'Rhythm Devils' segues into 'Throwing Stones', a track released on the album *In The Dark*. The video is a rapid montage of images of war juxtaposed against religious statues and stained glass windows which suggest hopes for spiritual transcendence. Like Nanci Griffith's version of Julie Gold's 'From a Distance', also released in 1987, 'Throwing Stones' takes the image of the Earth photographed from outer space as a basis for speculation on the futility of war. Men are 'staking turf / dividing up meat', while politicians are merely 'throwing stones'. The human race has a choice of leaving the Earth 'an empty stone / or a shining ball of blue we can call our home'.

Rock musicians concerned with environmental issues in the 1970s had to fight against the individualism of the 'Me Generation', which was reinforced by the egocentric tendencies of rock music. The songs of Don Henley both explored and reflected the contradictions of a culture whose motto was 'do your own thing'. Henley's environmentalist concerns gave an edge of conscience and intelligence to The Eagles. *Hotel California* (1976) mourned a country in moral decline; 'The Last Resort' closed the album by castigating the despoliation of California by developers, and the people's complicity in their greed and violence towards the land. The song criticises the American frontier mentality that has exploited the whole world to serve our 'endless needs'. Rich men have 'raped the land' merely to put up a lot of 'boxes', to which Henley's responds indignantly: 'Jesus, people bought 'em'. Yet the cool anger of the song turns into fatalism in the descending movement of the melody.

By 1990, Henley turned his environmental concerns into practical action by leading the successful campaign to save Walden Woods from development (Henley and Marsh 1991). This interest in conservation also features in 'Goodbye to A River' (2000), based on John Graves' memoir about the damming of a childhood river in Henley's home state of Texas (Graves 1960). The song is a moving elegy for a wild river destroyed by the hubris of human beings going 'where we ought not to go', and killing 'everything divine'. Yet while he rails against excessive development, Henley's most famous song as a solo artist remains 'The Boys of Summer' (1984), an exultant rock anthem for the consumerist thrills of American automobile culture, of driving with the 'top pulled down' and the 'radio on'. Henley's environmentalism is thus marked by the contradictions out of which rock music is constituted and finds its creative energies.

The extent to which popular music has actually inspired environmental activism is an empirical question beyond the scope of this book. While it would be naive to idealise and exaggerate the contribution made by rock music to environmental activism, it is also easy to be overly cynical and dismissive. Problems with rock star egotism apparently beset some of the environmental campaigns mentioned in this chapter. At the MUSE concerts against nuclear power, for example, many stars ran up large expenses and hall charges for performing too long, while Bernice Reagon of Sweet Honey and the Rock was upset that the concert took attention away from the mass protest in nearby

Battery Park, where a quarter of a million people were demonstrating against nuclear power (Denselow 1989: 178). Overall, though, popular musicians have at least given environmental causes valuable publicity. Noel Brown, head of the United Nations Environmental Program, commented that the involvement of The Grateful Dead in rainforest protection had brought 'more press attention to the issue than he had seen in the preceding ten years' (Denselow 1989: 230). Viewed in this light, the environmental activism of the musicians discussed here demonstrates a maturation of the social and environmental concerns first aired in the 1960s counterculture.

## 9.3 Anti-naturalism in American punk rock

The mid-seventies was dominated commercially by stadium rock and disco, while many of the musical acts popular in the sixties, some of whom we have encountered in this chapter, continued to sell records in large numbers. It was against this mainstream that American punk and new wave bands aggressively positioned themselves, rejecting what they saw as the blandness of stadium rock, exemplified by the Eagles and Fleetwood Mac, and the grandiose pretensions of British progressive rock acts such as Yes, Genesis and Emerson, Lake, and Palmer. Fans of the new music felt that the older generation of rock stars, from Mick Jagger to Bob Dylan, were too wealthy and remote to be in touch with the everyday realities of American life. American punk music did not attain the mainstream commercial success it did in Britain. Nevertheless, the ideological provocations of punk provided an important critique of American rock culture, and, from an ecocritical perspective, radically transformed musical representations of nature and the environment.

A product of big cities such as New York, Detroit and Cleveland, the American punk scene rejected the naturalistic values of sixties countercultural rock, which, as we saw in Chapter 7, manifested themselves in concerns for both the natural world and a 'natural' self which apparently only required liberation from social restraints to express itself fully. Punk, in contrast, had no interest in the natural world, and disdained both the back-to-the-land agrarianism of the hippie communards and the rural chic of the nouveau riche rock stars in their retreats in Laurel Canyon, Woodstock and Aspen. To the new punk sensibility, returning to nature was an impossible and evasive dream.

Punk's anti-naturalism extended to notions of the self. Unlike sixties countercultural rock, it took the self to be a totally artificial, social construction, and thereby embraced the provocations of 'camp'. In 'Notes on "Camp"' (1966), Susan Sontag had defined the 'camp sensibility' as a 'love of the unnatural: of artifice and exaggeration' (1967: 277). Camp sees no possibility of transcending what Marxists term 'reification'; that is, the way capitalism reduces human beings to things. Out of this failure of transcendence, camp formulates a survival strategy of revelling in kitsch, and creating perverse value out of the rejected, worthless junk of consumer capitalism.

Andy Warhol's Pop Art led the way to punk's fascination with camp. In the summer of 1966, when Warhol and his entourage, including the Velvet Underground, left New York on a visit to Los Angeles, he commented that 'vacant, vacuous Hollywood was everything I ever wanted to mould my life into' (Hoskyns 1996: 141). Warhol's sense of the self as an artificial construct, rather than a natural resource to be discovered and liberated, was later typified by punk, just as his rejection of personal expression and emotional affect became the adopted pose of what Richard Hell called the 'Blank Generation' of the mid- to late-1970s.

Film-maker and Warhol associate Paul Morrissey also enjoyed Los Angeles in 1966 precisely because it was an emblem of modernist alienation. 'LA I liked', he wrote, 'because the degenerates there all stay in their separate suburban houses. That's wonderful because it's so much more modern – people isolated from each other.' He added: 'I don't know where the hippies are getting these ideas to "retribalise" in the middle of the twentieth century' (Hoskyns 1996: 141). Morrissey may have had in mind the poet John Sinclair, onetime manager of the Detroit band the MC5, who believed that anarchist tribalisation was a solution to the way in which the industrial world enslaves people. 'People have their own houses', he wrote in his manifesto *Guitar Army* (1972), 'instead of living in tribal or clan groups. Isolated into families. And television is the ultimate low-energy trip. So our plan has always been to push high-energy music, high-energy life-style, bright clothes, weed, and communal living, which is because of its volatility, a high-energy thing. Get people into these forms and they relate to them. Because that's what people are supposed to be – free' (Sinclair 1972: 189).

For the punk sensibility, however, there could be no spiritual transcendence of commercial culture, and no return to a primitive or natural state of society. Radically disenchanted, punk rejected such utopian idealism, putting in its place parody and pastiche as ironic means of surviving the reifications of American consumer society. In songs such as 'Gimme Gimme Shock Treatment' and 'Suzy is a Headbanger', New York band The Ramones, as Bill Osgerby observes, both 'relished and lampooned' the images and stereotypes of teenage life they derived from the consumer society's 'fantasies of unbridled leisure, pleasure and carefree fun' (1999: 156).

The nihilistic side of this satire made punk less of a commercial proposition than sixties rock. The sixties counterculture, writes Guy Lawley, 'was arguably more easily assimilated by the mainstream than punk for many reasons. Though opposing "straight" society, it had some positive goals (e.g., ending both racist inequality and the Vietnam war); it preached flowers and love, so was prey to sentimentalisation; LSD and marijuana gave it a taste for fantasies and stories. Punk offered an angry, nihilistic creed and a doctrine of hatred. Much as the business sector tried to channel it into guises like New Wave and Power Pop, punk's negative credos were always likely to be commercial liabilities in any mass market' (Lawley 1999: 117). Yet the dystopianism and anti-naturalism of punk were also reminders that social realities could not be wished away. San Francisco artist Ken Sitz commented that the 'power of nihilism derives from its refusal of the status quo: deliberately *not* going for success or striving to "fit in". You look for the all that is *positive* in the refusal of constraints ... It's like a sleight-of-hand: you're negative, but for a *purpose*' (Vale and Juno 1994: 96). For songwriters such as Jello Biafra, the playful, blackly comic nihilism of punk, exemplified by the Sex Pistols' 'God Save the Queen', with its sneering refrain of 'no future', cleared the ground for political protest by getting rid of illusory hopes.

Biafra, born Eric Boucher, founded the Dead Kennedys in 1978 as a direct rebuttal to the country rock that dominated FM radio in his early youth in Aspen, Colorado. He saw the violence of his music as a necessary corrective to what he saw as the blandness of country rock and the post-sixties New Age culture it fed. 'I often wonder where psychedelia would have gone', he commented, 'if people like George Harrison had gotten into Japanese kabuki music with its overtones of pain and violence, instead of Indian music which is more blissful' (Vale and Juno 1994: 12). In the Dead Kennedys' 'California Ubër Alles' (1980), Democratic Governor Jerry Brown outlines his plans for California: suede and denim, meditation, jogging and Zen Buddhism are all to be enforced by the state mind-police. Biafra scandalously compares California to Hitler's Germany, by extending his attack on New Age fads to the cult of organicism that also underpinned Nazism. The 'Zen fascists' who seek total control of California are '100% natural'; Brown tells his audience they will be forced to wear a 'pretty flower' on their clothes, or else will be executed with 'organic poison gas'.

Biafra's delight in bad taste was a tactic of confrontation aimed at alerting people to the class divisions and decaying social conditions in post-industrial America. Musically, punk's harmonic and melodic simplicity, basic instrumentation and embrace of dissonance and noise (in the sense of both unpitched sound and high volume) connoted this social collapse, while its cover art replaced the benign nature imagery of sixties countercultural rock with dirty realist reminders of urban decay. The front cover of Pere Ubu's *The Modern Dance* (1977) depicted a grim, monochrome cityscape, in which a heroic male worker waves a flag, as if carrying the last hopes for a socialist revolution. The back cover photograph showed smoke billowing out of industrial chimneys in the band's hometown of Cleveland, Ohio. Unlike the Jefferson Airplane's *After Bathing at Baxter's*, there is no quaint biplane to carry the hippie elite away from the urban squalor, only a ghostly white jet aircraft flying threateningly low over the city.

Pere Ubu's music created a mood of post-industrial decay, as David Thomas' pinched, anguished vocals howled over aggressive guitar and drums, wailing horns and disconcerting *musique concrète* sound effects. 'Non-Alignment Pact', the opening track of *The Modern Dance*, begins with piercing electronic shrieking, as a love relationship is reduced to the brutalities of Cold War *realpolitik*. The title track includes sounds of people laughing in a dance of denial. 'Laughing' searches for possibilities of survival in the 'empty spaces of this life', under an apocalyptic sky in which the 'stars are coming all undone'. The cold wind effects that blow through 'Street Waves' may suggest a nuclear winter, as Thomas' garbled vocals alienate the listener, despite the upbeat rock rhythm. 'Chinese Radiation' picks up the reference to Maoism on the album cover; panicked vocals and crowd cheers break down into a mechanistic, affectless refrain ('she'll wave a red book') over fractured piano chords, as utopian political hopes die a lingering death. For Pere

Ubu, resistance to such apocalyptic despair emerged from the manic humour and exaggerated grotesquerie of Thomas' vocal performance, which worked as both social critique and emotional release.

As punk rock expanded stylistically into 'new wave', the natural world, absent from the post-apocalyptic urban landscapes evoked by records such as *The Modern Dance*, began to re-emerge. Indeed, as his songwriting developed, David Thomas himself began to include images of the natural world to represent a resource of escape from the oppressions of urban modernity. In 'The Birds Are Good Ideas', on his solo album *The Sound of the Sand and Other Songs of the Pedestrian* (1981), people look sadly at their shoes until they hear the 'happiness' expressed in birdsong. The music is a deranged mixture of rockabilly guitar, Latin-American trumpet and syncopated dance rhythms. Thomas' manic humour is both child-like and childish; a desperate and comically forced attempt to find hope and joy in the natural world. In his version of The Beach Boy's 'Surf's Up' (2001), the gentle pastorale of 'muted trumpeter swans' and 'dove nested towers' begins as genuine utopian longing, but the hope is gradually undercut by harsh electronic timbres, disturbing in their artificiality. In Thomas' songs, then, a complex pastoral measures the distance between an ideal of pristine nature and a post-industrial reality that cannot easily be transcended or transformed.

David Thomas' turn to the natural world for emotional and spiritual release may be seen as a counter-tendency within American new wave music that ultimately continued, rather than rejected, the utopian promise of the sixties rock counterculture. Romantic primitivism also lived on in Patti Smith's 'Land' (1977), which enacted a Dionysian ritual of transgressive female sexual desire. The epic ride on a 'mare black and shining with yellow hair' enters an immersive 'sea of possibility' that brings together African-American dance cultures (the pony, the twist and the watusi), Chuck Berry's Johnny B. Goode and the French nineteenth-century Symbolist poet Arthur Rimbaud. In Rimbaud's oft-quoted phrase, a poet 'makes himself a *seer* by a long, immense and reasoned *disordering* of *all the senses*' (italics in the original) (Rimbaud 2005: 238-9). Smith's visionary journey finds spiritual and physical renewal through immersion in the natural energies of the unfettered horse and the engulfing sea.

In contrast to this return to nature in David Thomas and Patti Smith, the Talking Heads playfully questioned conceptions of nature and authenticity in a way typical of the postmodernist aesthetics celebrated by some critics in the 'post-punk' music of the 1980s. '(Nothing But) Flowers' (2006) is postmodernist in its ironic use of pastiche and its play with codes of signification. David Byrne's lyrics play on the idea of technology as second nature, as if Joni Mitchell's 'tree museum' in 'Big Yellow Taxi' is better than a real forest. In a society of the near future, urban spaces have reverted to wilderness, and the faux-naif narrator longs for the developed, commodified landscape of twentieth-century America. The land 'used to be real estate', he laments, 'Now it's only fields and trees'. 'If this is paradise', he whines, 'I wish I had a lawnmower'. Byrne's ironic complaint is set to the lively rumba rhythms and bubbling treble guitar arpeggios of West African soukous. This witty musical pastiche leaves open the question of which Eden is to be preferred, the natural one signified by African music, or the urban landscape of 'honky-tonks, Dairy Queens, and 7-Elevens' celebrated in the lyrics.

## 9.4 Indie rock: a return to 'nature'

The end of the seventies saw declining record sales in the United States, a trend reversed not by punk rock or new wave, but by the rise of the compact disc and the founding of MTV in 1981. Music television reinforced the socialisation of young people into capitalist values, and further defused the dreams of the sixties counterculture for a unified and radical youth culture. As David Buxton argues, 'By ignoring the possibilities of an alternative organisation of culture in favour of symbolic radicalism, rock music helped play an important role in the production of the necessary creative imaginary for capitalist reorganisation' (1990: 435).

The 1980s saw a proliferation of marketing categories that reflected the fragmentation of audiences for popular music. For the rock audience of this period, the aura of countercultural authenticity was adopted by 'indie' rock, which adheres to the idea, as Eric Weisbard puts it, that 'one could be a consumer without the traditional associations of gross commodification, audience passivity, and massness' (Gracyk 1996: 151). Yet most 'indie' labels are funded by the major companies as either niche markets or as research and development spaces in which new talent can be nurtured. Moreover, bands such as R.E.M. followed the predictable career path when they moved from independent label IRS to Warner Bros. in 1989.

Nevertheless, the indie scene, however broadly defined, has been a creative space in which an environmental politics has emerged. In the 1980s, some American indie rock bands paralleled the example of British musicians such as Joe Strummer, Paul Weller and Billy Bragg, who had steered British punk rock away from the blackly comic nihilism of the Sex Pistols into an older tradition of left-wing, proletarian protest derived from the folk music of A. L. Lloyd and Ewan MacColl. If the overriding desire of the rock ego is solipsistic, the collectivist element in the folk tradition enables concerns for the social to emerge. By drawing on folk and country traditions, American indie rock has thus restored to rock music a concern with the natural world after the anti-naturalist provocations of punk in the mid-1970s.

Michael Stipe's lyrics to R.E.M's 'Cuyahoga' (1986) are a collage of three different narratives formed around the Cuyahoga River in Ohio: a childhood memory of grazing his knee when swimming in a river with a friend, the Indian tribes who 'hunted, danced and sang' near the river in the distant past, and the apocalyptic events of June 1969, when 'we burned the river down'. These images are linked by a sense of violation, guilt and loss, whose emblem is the spilling of blood in the river. The river is now 'gone', existing only in memories and tourist photographs; yet there is an optimistic, resistive call to 'start a new country up'. Repetition gives the lyrics the emotional impact of a lament, while the dynamic shadings of the folk-rock accompaniment, led by Peter Buck's mandolin, is at various times angry and contemplative.

The Pixies' 'Monkey's Gone to Heaven' (1989) alludes to environmental threats in a more solipsistic manner. In the first verse, an 'underwater guy who controlled the sea' has been 'killed by ten million pounds of sludge / from New York and New Jersey'. The suggestion of punishment for hubristic attitudes to nature continues in the next verse, when 'everything is gonna burn', including the complicit narrator: 'I'll get mine too'. The last verse obliquely references numerology and Manicheanism: 'if man is 5', then 'the devil is 6' and 'god is 7'. The universe is ruled by occult forces, and the refrain – 'this monkey's gone to heaven' – returns to the notion of hubris: the monkeys used in early space flights are an emblem of the failure of human evolutionary progress. The song thus presents fears of environmental apocalypse as a confusing, uncontrollable nightmare, both technological and mystical, and thereby without the uplifting appeal for social change that R.E.M's 'Cuyahoga' gained from its proximity to the proletarian realist folk tradition.

American indie rock entered the twenty-first century by continuing to draw on a wide range of musical influences, from country to electronica, to explore contemporary experience in an environment heavily mediated by new electronic technologies. In Grandaddy's 'I'm on Standby' (2003), the narrator has been reduced to a computer 'powered down for redesign'. Digital technology has become second nature, and acceptance of this condition, and the lack of emotional affect it implies, seems inevitable. However, in 'The Group Who Couldn't Say', a generation of computer nerds is taught the pleasures of contact with the natural world. The song tells the ironic story of a pop group whose record company rewards them for being the 'shrewdest unit movers' by giving them a trip to the countryside. Exposure to the natural world outside the modern computerised recording studio proves to be an experience of wonder beyond words. Becky learns that her 'drag and click' had 'never yielded anything as perfect as a dragonfly'. Musically, the pastiche of Beatles-style harmonies evokes both the lyrical beauties of nature and a sense of ironic, cool distance from them.

Natalie Merchant has written a number of perceptive songs of environmental concern, in both her new wave band 10,000 Maniacs and her subsequent solo career. 'Poison in the Well', released on 1989's *Blind Man's Zoo*, is a dramatic monologue in which a man gradually realises that he and his family have been lied to about the contamination of their water supply. The official euphemisms – 'someone's been a bit untidy and there's been a small spill' – have failed to overcome his anxieties. The optimistic jauntiness of the music contrasts ironically with the fears expressed in Merchant's mournful vocals.

Musically, Merchant's solo album *Motherland* (2001) expanded into blues and reggae, which provided a contemplative space for her continuing exploration of environmental issues. The title track is a slow, melancholy complaint against concrete sprawl driven by 'the lust and the avarice, the bottomless, the cavernous greed'. The song offers no solution; instead, the chorus appeals to the motherland to 'lullaby me to sleep' in a tone of childlike vulnerability and fatalism.

The indie folk-rock scene, which includes Natalie Merchant, has also encouraged songwriters such as Ani di Franco and Dar Williams, whose songs similarly move outward beyond diaristic self-exploration towards social and environmentalist concerns. In this way, the 'indie' scene overlaps with the re-emergence of politically committed singer-songwriters working more directly in the folk tradition, such as David Rovics and Tish Hinojosa, whom we encountered in chapter 6.

## 9.5 African-American hip hop and environmental protest

Hip hop culture emerged amongst black and Hispanic youth in New York City in the late 1970s, as people of colour suffered disproportionately from the urban 'renewal' policies of Robert Moses which destroyed old communities in the South Bronx (Berman 1982). Tricia Rose emphasises the influence of this post-industrial urban environment on the development of hip hop. Rap and graffiti, she writes, are expressions of popular protest marked by 'especially aggressive public displays of counterpresence and voice. Each asserted the right to write – to inscribe one's identity on an environment that seemed Teflon resistant to its young people of colour; an environment that made legitimate avenues for material and social participation inaccessible' (Rose 1994: 59-60).

Rose argues that hip hop music enacts political resistance at a formal level. Stylistically, hip hop's main features of flow, layering, and rupture 'simultaneously reflect and contest the social roles open to urban inner-city youths at the end of the twentieth century' (22). These techniques are a 'blueprint for social resistance and affirmation: create sustaining narratives, accumulate them, layer, embellish, and transform them. However, be also prepared for rupture, find pleasure in it, in fact, *plan on* social rupture. When these ruptures occur, use them in creative ways that will prepare you for a future in which survival will demand a sudden shift in ground tactics' (39).

Although she does not explicitly bring out the environmentalist implications of hip hop's origins in the post-industrial city of late 1970s America, there are hints in Rose's work that hip hop has eco-aesthetic implications at the level of both form and content. Sampling and scratching, for example, can be seen as forms of musical recycling. 'Worked out on the rusting urban core as a playground', Rose observes, 'hip hop transforms stray technological parts intended for cultural and industrial trash heaps into sources of pleasure and power' (22). Drawing on Houston Baker's work on the blues, she argues that hip hop DJs are using 'the tools of obsolete industrial technology to traverse contemporary crossroads of lack and desire in urban Afrodiasporica communities' (35). While this argument neglects the contribution of state-of-the-art studio technologies to the production of the complexly layered soundscapes of many hip hop recordings, the ready-made, bricolage aspects of hip hop are nevertheless important.

Adam Krims identifies four subgenres within rap: 'party', 'mack', 'jazz/bohemian' and 'reality'. Rappers who have voiced explicit concerns about environmental issues tend to come from the latter two genres, or, as Krims writes of the rapper Common, from a position on a 'cusp' between the two (Krims 2000: 70). Krims defines the 'reality' genre as 'any rap that undertakes the project of realism, in the classical sense, which in this context would amount to an epistemological/ontological project to map the realities of (usually) black inner-city life' (70). It includes the subgenre of gangsta rap, which emerged from a Los Angeles which, like New York City under Robert Moses, saw working-class African-American and Hispanic communities destroyed by urban redevelopment. Mike Davis shows how, at the end of the 1960s, the disenfranchisement of people of colour contributed to the rise of gang culture in the city (1990: 293-300). Eithne Quinn sees gangsta rap culture as both a survivalist and an entrepreneurial response to this decaying urban environment. It was, she writes, a way for black people to adapt to hostile living conditions through an image of 'desultory, resourceful enterprise, of playing the system whenever and wherever possible in an age of profit – and market – driven restructuring of economic life' (Quinn 2005: 43). Like Hollywood movies such as Warners' *New Jack City* (1991), gangsta rap trades in the mixed message of the gangster's rise and fall. The lure of the 'thug' life of romantic outlawry has been reinforced by a music industry keen to cash in on sensationalist and exotic images of black life. From an ecocritical perspective, gangsta rap tends to neglect issues of environmental justice, but for understandable reasons.

In addition to gangsta rap, Krims' category of reality rap also includes 'knowledge' rap, which covers 'political or historical tales, religious lessons, or other didactic modes of information' (2000: 79). This subgenre emerged both from the traditionally didactic role of African-American oral traditions and from contemporary commercial factors. 'In particular', Krims writes, 'there is a certain subset of rap fans for whom the music bears a mission of spiritual/cultural elevation. That phenomenon itself is inseparable from an end-of-the-twentieth-century development of Afrocentric thought, including a serious investment in the possibilities of rap music for communication and social transformation' (80). Musically, reality rap displays a 'hardness' signified by a dissonant layering of sonic materials (73).

Grandmaster Flash and the Furious Five's 'The Message' (1982) is the early classic of reality or knowledge rap which offered an explicit engagement with topical issues of social, if not explicitly environmental, justice. The nervous, stuttering syncopation of MC Melle Mel's vocal refrain enacts the effects of urban poverty on black male subjectivity: 'don't push me cos' I'm close to the edge / It's like a jungle sometimes / It makes me wonder how I keep from goin' under'. The narrator gives quick character sketches of the people in his immediate environment: the bag lady who has turned to prostitution to survive, his brother addicted to daytime television, his son who is reluctant to go to school, a girl pushed under a subway train, and the 'thugs, pimps and pushers' familiar to 'ghetto' songs since the funk era of the early 1970s. The list ends with a direct address to a young suicide in a prison cell. The rap accuses failed institutions, from the schools to the police, and, in keeping with African-American music from the blues onwards, understands the causes of social collapse as fundamentally economic: in an era of 'double-digit inflation', 'it's all about money, ain't a damn thing funny'. Education is held up as a possible way out of the ghetto, a hope underlined by the internal rhyme: 'You say I'm cool, huh, I'm no fool / But then you wind up droppin' out of high school'. Despite the promise of its title, however, the song offers no overt political solution. Instead, it is a relentless testimony of urban disaster; despair is offset by the catchy synthesizer riff and steady bass groove, which provides an element of utopian uplift and release.

Krims' category of 'jazz/bohemian' rap shares the didactic function of reality rap, but is aimed at a different audience. 'The topics addressed in jazz/bohemian songs', he argues, 'tend to explain the frequent claims of its being a 'conscious' genre, in contrast to the purported limited political vision of other rap genres'; he adds, however, that all hip hop genres show 'virtually unrelenting stylistic and historical self-consciousness'. The jazz/bohemian genre, though, 'emphasises the value of gaining knowledge, criticises the assumed "negativity" of some other rap genres, or perhaps promotes Afrocentric values and artistic traditions' (68). Musically, it favours an upbeat, pop feel, and makes references to soul and pop styles from the 1950s to the 1980s.

Mos Def may be described as a rapper on the cusp between 'reality' and 'jazz-bohemian' styles. 'New World Water', on *Black on Both Sides* (1999), explores the class politics of water scarcity in a world where 'every drop counts'. In this apocalyptic anti-pastoral, Nature is getting revenge for its mistreatment at the hands of white people: 'Fools done upset the Old Man River / Made him carry slave ships and fed him dead nigga'. As the pun in the title suggests, the scope of Mos Def's critique is global: Americans are wasting water 'on some leisure shit / And other nations be desperately seeking it'. As in 'The Message', class inequalities are the basis of all social problems: 'The type of cats who pollute the whole shore line / Have it purified, sell it for a dollar twenty-five'. The closing refrain exposes the economic message, in a comic Cockney gangster accent: 'it's all about getting that cash (money)'. With 'New World Water', rap extends its range of concerns to an explicit politics of environmental justice conceived in terms of race, class and nationhood.

Jazz-bohemian rap has revived a tradition of African-American pastoral which, as we will see in Chapter 12, is also a prominent feature in jazz. Arrested Development and Michael Franti's Spearhead combine Afrocentric imagery with 'positivity', or faith in the power of positive thinking. Like Marvin Gaye and Stevie Wonder, Arrested Development dedicated their album *3 Years, 5 Months and 2 Days in the Life Of ...* (1992) to God: 'Lord – through Arrested Development may your work be done and the vibes created from these sound recordings positively effect (sic) every living entity in the entire universe'. In 'Children Play With Earth', biophilia is redemptive. The song is an anthemic appeal to 'African' children to reject European acculturation: 'dig your hands in the dirt' rather than play with Nintendos and watch television. Direct contact with the soil, playing in the grass, climbing trees and eating fresh fruit will restore bodily and spiritual health, and heighten attentiveness to the natural world: 'watch and listen to the neglected mother of all'. The didacticism of the lyrics is tempered by the energetic vocal delivery and the joyfully dissonant layering of sampled drums, reggae horns and funky, slapped bass. However, the same objections that some listeners have made to protest folk music have also been made to jazz-bohemian rap. For fans of reality rap such as the critic Peter Shapiro, the didacticism of Arrested Development was merely 'feel-goodism and liberal guilt-tripping of the smuggest, most pompous, insufferable sort' (2005: 15).

Like Arrested Development, Michael Franti draws on pastoral and organic imagery for his persona of hippie-Rasta child of nature: 'you could walk a mile in my bare feet'. In 'What I Be', on *Everyone Deserves Music* (2003), the African-American self imaginatively projects itself into nature, becoming mountains, sky, seed, trees, leaves and oceans: 'If I could be a seed, I would give birth to redwoods trees and / If I were the trees, I'd generate the freshest air to breathe in'. The song ends with the ultimate merging of subject and object, in a way that recalls Warren Fox's concept of transpersonal ecology: 'If I could be you, you could be me' (Fox 1990). The slow tempo gives space for decorative vocalese, bright electric guitar fills, synthesized flutes and bells. Jamaican dub-style echo makes for an ambient, immersive sound that matches the cosmic consciousness of the lyrics.

Elsewhere on the album, Franti's environmental concerns extend to a critique of total war, while his musical eclecticism extends into punk rock. 'We Don't Stop' is an anti-war tirade which uses insistent, biblical repetition: there is a 'war for oil', a 'war on birds and a war on bees / They got a war on hippies tryin' to save the trees'. The Clash-like chorus references punk in its celebration of the righteous anger of 'rebel rock'.

As Tricia Rose observes, hip hop was never outside of the commercial sphere: 'The contexts for creation in hip hop were never fully outside or in opposition to commodities; they involved struggles over public space and access to commodified materials, equipment, and products of economic viability' (1994: 40). Indeed, hip hop became a mainstay of programming on MTV in the 1990s. In 2006, the video for 'Trees' by Dr. Octagon was shown as part of station's THINK campaign, which advocates global citizenship and activist concern. 'Trees' is a pop rap song which musically references 1980s synthesizer pop music. Like Arrested Development's 'Children Play With Earth', it is addressed to a younger audience than Mos Def's 'New World Water', and lacks his racial and class analysis. The lyrics playfully combine irrational connections with apocalyptic references: 'A bunch of chemicals y'all turning it loose. Carbon dioxide comin' out to you. Trees may be extinct like the elephants'. The video shows schoolchildren playing around in a science laboratory; the carnivalesque visual humour recalls George Clinton and Public Enemy's Flavor Flav. The satire on the authority of science hints at conspiracy: 'Information is more concealed'.

Dr. Octagon's 'Trees', like Michael Jackson's 'Earth Song', shows how, at the beginning of the twenty-first century, environmental protest has been incorporated by global media corporations such as MTV, as long as it tends to be vague, non-specific and entertaining. Writing of the Live Aid charity records 'Do They Know It's Christmas'(1984) and 'We Are the World' (1985), which were heavily featured on MTV, Jack Banks concludes that 'music videos that receive broad distribution and exposure on U.S. media systems only seem able to grapple with social problems like poverty in the most apolitical, inoffensive manner possible' (1996: 205). The THINK campaign, from this viewpoint, is MTV's public conscience, which does not affect the station's overall concern for commodity consumption. 'Bourgeois institutions', writes Marxist geographer David Harvey, 'have a long history of exercising "repressive tolerance" and the current state of environmental/ecological debate, over goals, values and requirements, appears more and more as an excellent case study of how a limited articulation of difference can play exactly such a sustaining role for hegemonic and centralised control of the key institutional and material practices that really matter for the perpetuation of capitalist social and power relations' (1996: 174-5).

Yet the processes of incorporation at work in the corporate music industry are neither total nor simply deterministic. The slogan of the THINK campaign is 'Your Cause, Your Effect', and the web-site emphasises both individual lifestyle change and support for fair trade organisations such as EarthShare. By providing opportunities for social networking, blogging and the uploading of home videos, MTV's THINK provides a wide, potentially politicised context for the act of consuming popular music which is the corporation's main source of income: the consumer can no longer be conceived of as simply passive, but as a potentially active participant in cultural and political change.

The problem with the environmental politics of popular music at the start of the twenty-first century thus remains the same as at the end of the sixties: the realisation that music cannot save the planet on its own. As Alistair McCulloch noted, the 'green paradox' of the globally televised Live Earth concerts in July 2007 was that 'the performers whose stated aim is to encourage others to consume less and act in a "greener" manner, tend to be both conspicuous consumers and leave environmental footprints much larger than the people who were watching them [...] The celebrity lifestyle is at odds with environmental sensitivity' (2007: 1). Comedian Chris Rock summed up the limitations of popular music as an agent for global environmental change from the stage of the Live Earth concert in London: 'I pray that this event ends global warming the same way that Live Aid ended world hunger' (Collett-White 2007: 2). But responses to the current political impasse can either be cynicism and resignation or activist empowerment and resistance. The songs discussed in this chapter suggest that American popular music, from rock to hip hop, is encouraging the latter as much as the former.

# Chapter Ten

# World music

The term 'world music' was first coined by ethnomusicologists in the early 1970s to describe their research into the world's folk musics. In 1987, however, it was adopted as a marketing category by the British music industry to distinguish the new, hybrid forms of popular music which were emerging from countries outside the Anglo-American sphere from the indigenous folk forms on which they were based. In the United States, these musical fusions are sometimes referred to as 'world beat', to set them apart from non-Western folk musics, to which the earlier term 'world music' is still applied. Increasingly, however, both the music industry and its audience are using the latter term to apply to both forms of music. In what follows, the term 'world music' is used in the sense of 'world beat', to designate the new hybrid forms of popular music, as the chapter explores the discourses of ecology and environmental concern that have grown around the reception of such music since the 1980s (Taylor 1997: 3-4).

## 10.1 World music and globalisation

In *Born to Groove*, musicologist Charles Keil has formulated an explicitly ecological theory of music. Educating children in drumming and dance, he argues, will foster a concern for both cultural and biodiversity, and thereby help them develop a more benign relationship with the natural world than that practiced by earlier generations. He outlines his utopian program for what he calls the 'the oncoming economically sustainable and paleoconservative centuries' as follows: 'If we help each child to feel the pleasure of "groove locally", creating, being in flow, finding focus, capable of being completely individuated and completely part of a local social whole at one and the same moment, then we can "cool globally" – at least slow the loss of species and cultural diversity'. Mass participation in music-making may produce a society less addicted to the idea that commodity consumption is the main means to happiness. It may even end war. 'The logic', Keil writes, 'is impeccable: happy people, people experiencing pleasure and satisfaction in their creativity, will not crave the worst forms of power-over, will not want to participate in empires and "administrative massacres" (Arendt) or in terrorism' (Keil and Campbell: 2006: 9.1).

The conceptual link that Keil makes between music and biodiversity derives from the notion that many organisms communicate with their environment through sound. Music-making therefore restores our connection with other living species more than any other form of cultural activity. 'Did hominids or humans take a wrong evolutionary turn', he asks, 'when we abandoned echolocation for "culture?" When we gave up sound minds in sounds bodies to look at books, TV and computer screens? Could music-dance-ritual be our version of echo-location and the key to resuming our co-evolution with nature?' (Keil and Campbell: 2006: 11.1). Although Keil criticises the alienating effects of mass media technologies, his argument is nuanced enough to recognise the enabling role they can also play in cultural production. It is for him a question of the degree to which experiences of music are passive and mediated, or active and participative.

Of course, Keil overstates the power of music to do good in and of itself, and probably underestimates the lack of competitiveness and egotism amongst some musicians themselves. But his understanding of the healing powers of music is nevertheless important. Musicologist Martin Stokes writes of the mood of 'gentle millennial mysticism' which surrounded the emergence of world music at the end of the twentieth century, which seemed to presage a new global future beyond the nation-state. 'What better image of this world', he explains, 'than the grooving together of cultures remote from one another in music, that perennial source of social warmth in modernity's cold world?' (Stokes 2003: 297). As Stokes rightly points out, such idealistic notions tend to mask the complex realities of power and control that make world music a highly contested terrain. But this critical argument does not discredit Keil's faith in music as a source of 'social warmth': his utopian theory is, in Ernst Bloch's terms, concrete enough to recognise the vital cognitive, physiological and social role that artistic creativity can play in everyday life, and the way this can be extended to concerns for wider social and political matters, including the health of the environment.

The idea that music is a means to ecological interconnectedness is itself a product of the economic and cultural globalisation that has accelerated since the 1960s, especially with the popularisation of information technology and jet air travel. For some members of the new middle-class which runs the information society, the new global electronic technologies are fostering a 'One-World' vision of quasi-mystical togetherness. When Stewart Brand founded the *Whole Earth Catalogue* in 1968, he put on the cover the first photograph of the Earth viewed from outer space, taken by the Apollo 8 astronauts. This image of the Whole Earth prompted public debates about the need to care for the health of the entire planet (Garrard 2004: 160ff). As geographer David Smith notes, the modern media can create 'a sense of involvement in distant lives' that can in turn promote 'an extension of a sense of responsibility' for those lives (Buell 2001: 73). For all their political limitations, the Live Aid, Live8 and Live Earth concerts are manifestations of such globalised concerns.

Yet the One World concept tends to mask the social and political realities of globalisation which have seen an intensification of damaging social and environmental conflicts (Robertson 1992: 8-10). Moreover, such McLuhanite notions of Global Village are also being deployed as marketing strategies, both by small New Age entrepreneurs and by big businesses like Nike and Benetton. In the music industry, the new ideology of diversity within global unity reflects the globalisation of the industry itself in the 1980s, and masks its transformation into an oligopoly. By 2009, after a series of horizontal and vertical mergers and acquisitions, four major labels, Universal Music Group, Sony Music Entertainment, EMI and Warner Music Group, control most of the global market for legally recorded music.

Stylistically, the rise of world music reflects the complex dialectic of homogenisation and heterogenisation that has resulted from globalisation, understood both as the internationalisation of the mass media and the increased movement of peoples across different nations. In one respect, the capitalist drive by the transnational record companies to find international sources of profit has led to stylistic homogenisation, or an Americanisation of popular music, so that hip hop, heavy metal and electronic dance music are now global styles found in cities all over the world. Yet globalisation has also led to the reassertion and reinvention of local styles and traditions. As ethnomusicologist Philip Bohlman writes, musicians in the new 'global' cities of New York,

London, Paris and Rio de Janeiro 'do not just perform in public or private settings, nor do they function within single ethnic communities or class structures, but rather they form complex affiliations that cut across socioeconomic, religious and ethnic boundaries' (2002: 136). Accordingly, as Simon Frith argues, a simplistic, top-down concept of globalisation in discussions of world music, 'with its intimations of the inexorable forces of history and/or capital', should be replaced by 'an understanding of networks – globalisation from below, as it were' (2000: 319).

The organic metaphor of 'hybridity', introduced into music scholarship in the 1990s, is intended to map these complex processes of cultural exchange, in place of the older critical model of 'cultural imperialism', which implied the simple imposition of economic and social processes from the top-down. David Toop writes that the idea of 'roots' music, another commonly used organic metaphor, is 'improbably, still fixed deeply in the soil as a point of origin despite the passing of years and the invasive ubiquity of mass media and consumerism' (2004: 177). Deleuze and Guattari's 'rhizome' provides another organic model to challenge the 'arborescent' assumptions of the idea of 'roots', that risk viewing musical cultures in terms of fixity and stasis. Instead, hybridity and 'rhizomatics' map the openness to adaptation and change that world music embodies.

Hybridity is also the key to Charles Keil's idea that world music, by drawing on indigenous folk musics, can rejuvenate 'Western' culture to the point of encouraging concerns for the natural world. Musicologist Timothy Taylor lists the main discourses used to describe world music by the music industry, musicians and audiences alike: 'rejuvenation, novelty, authenticity, originality, the "real", and the spiritual' (1997: 14). Of particular relevance to the discourse of ecology that has emerged from world music is the way audiences associate authenticity with the premodern, as they demand, as Taylor puts it, 'some discernible connection to the timeless, the ancient, the primal, the pure, the chthonic; that is what they want to buy, since their own world is often conceived as ephemeral, new, artificial, and corrupt' (26). This notion of authenticity as 'emotionality and spirituality' involves the 'spiritualisation of decontextualised and deritualised musics' (25).

In order to discuss the environmentalist implications of this decontextualisation of tradition in world music, we will look at two examples. Merl Saunders released *Blues from the Rainforest* in 1990 to support the Rainforest Action Network. Described in Saunders' liner notes as 'environmental earth music', the album is a collaboration with Serbia-born drummer Muruga (formerly Steve Booker) and Jerry Garcia. The musicians explore a relaxed, New Age jazz groove which evokes an arcadian natural world as lush, spacious and bright as the production. On the title track, Garcia adds delicately meandering lead guitar over Saunders' gentle synthesizer *ostinati*, while Muruga's shaker and cymbals blend into ambient recordings of rain and bird calls. The music seamlessly blends traditional African rhythms with electric guitar and synthesizers. This creative reconciliation of the old and the new, the spiritual and the material, is at the heart of the ideology of world music, and is illustrated in two of the sources whom Saunders acknowledges in his liner notes: Muruga, whose 'belief in Eastern philosophies', particularly yoga and meditation, provides the basis for his interest in the indigenous rhythms that structure the music, and sound engineer Bill Thompson, Saunders' 'MIDIcian and programmer', who is thanked for 'helping develop new sounds on my synthesizers, and forcing me to take that next step into advanced technology'.

Mickey Hart's Grammy-award winning *Planet Drum* (1991) brought together virtuoso drummers and percussionists from around the world, including Airto Moreira from Brazil, Babatunde Olatunji and Sikiru Adepoju from Nigeria, Zakir Hussain from North India and T.H. Vinayakram from South India, is another act of self-conscious hybridity and fusion that linked world music with global environmental awareness. Like *Blues from the Rainforest*, *Planet Drum* also celebrates a fusion of old and new. The musicians play a wide range of traditional percussion instruments as well as new ones created especially for the recording, and some which, as Hart puts it, were 'a combination, using new technology to enhance known sounds'. His technical notes lovingly list the state-of-the-art recording technologies used in the project. Yet many of the titles on the album invoke the role of drumming in the rituals of hunter-gatherer cultures: 'Dance of the Hunter's Fire', 'The Hunt', 'Temple Caves' and 'The Dancing Sorcerer'. The album thus suggests the idea of music as primal ritual, while at the same time presenting that music in a deritualised form. As in the sixties rock scene of which Hart, as drummer with The Grateful Dead, played an important part (see Chapter 7), *Planet Drum* thereby attempted to reconcile ideas of primal nature with new electronic technologies in a way that the musicians themselves found entirely unproblematic.

Indeed, in the book that accompanied the project, Hart emphasised the sacredness of music: 'Our word *religion* comes from the Latin and means "to bind together". A successful religion is one that binds together all the fundamental rhythms that each of us experiences: the personal rhythm of the human body, the larger social rhythm of the family, tribe, or nation, and the enveloping cosmic rhythms of the planet and universe. If a religion "works", its followers are rewarded by a new dimension of rhythm and time – the sacred.' The notion of rhythmical entrainment, which we discussed in Chapter 4, is thus central to the place that Hart assigns to music in ecological thought. 'Because we are a part of nature', he writes, 'it is likely that we are entrained with the larger planetary and universal rhythms that surround us' (Hart and Lieberman 1991: 17).

The closing track on *Planet Drum* epitomises the ecological concerns of Hart's project. 'Mysterious Island' blends Brazilian jazz singer Flora Purim's swooping, seagull-like vocalisations with recordings of ocean waves in Hawaii, bird whistles, wind chimes, tambourine and a gentle drum groove. The piece thus explores resonances between the human body, electronic and acoustic technologies and the natural world, by mixing, as Hart puts it, 'the natural elements of water, rain, blowing wind, and birds with the sound of metal bells and the human voice'. In so doing, it may be seen enacting Charles Keil's idea that concerns for biodiversity and cultural diversity may be part of the same ecological vision.

Although Hart said that the album had 'no agenda other than to play and see what rhythms emerged', he nevertheless dedicated it to the Planet Drum Foundation in San Francisco, the bioregionalist organisation set up in 1972 by, amongst others, former Digger Peter Berg, whom Hart knew from his days in the Haight-Ashbury in the late sixties. He thereby implicitly linked the hybridity of world music to a bioregionalist politics of locality, place and respect for non-human nature. This is the social, political and environmentalist dimension of the notion of the sacred which Hart sees as fundamental to drumming as a cultural practice.

Although Timothy Taylor finds the decontextualisation of traditional musics a problematic aspect of world music, musicians such as Mickey Hart clearly do not consider it an obstacle to the sense of environmental awareness they desire from their music. Moreover, the appeal to the sacred in these New Age ideas of music has been encouraged, rather than made more difficult, by the mechanical and electronic reproduction of music. As Evan Eisenberg argues, ritual functions have always surrounded the collecting of, and listening to, recorded music. Indeed, even though recording technology tends to decontextualise sacred musics from around the world, the act of playing recorded music nevertheless tends to restore a sense of sacred ritual. 'With the phonograph', Eisenberg writes, 'music of every description – sacred, secular, art, entertainment, scavenged from cultures present and past – is injected into everyday life. In terms of context nothing is sacred; all music is secular. But in its manner of performance all music is sacred, because the phonograph always plays it exactly the same way' (1987: 64). The same may be said of the IPod, and the ritual pleasures of searching for and downloading music on the Internet. Rather than destroying the aura that surrounds the consumption of art, as Walter Benjamin suggested, mechanic and electronic reproduction have reinforced it (1973: 223).

A greater obstacle to music's potential for ecological connectedness is surely the individualisation of listening encouraged by the mechanical and electronic reproduction of music. Listening to recorded music through headphones in a public space, or alone in a room, is an individualistic experience that reinforces a sense of privacy. As Eisenberg puts it, 'Private music, music no one else hears, reinforces your invisibility by giving you a private structure of time to set against public time' (1987: 27). Yet New Age theories of music and ecology want to reclaim collective modes of listening from such atomistic individualism. Hence the enthusiastic embrace of McLuhan's theory of the Global Village, with its vision of electronic technologies fostering new forms of collective, democratic and participative culture.

This ideal necessarily includes virtual, as well as face-to-face, communities. Since its popularisation in the 1920s, radio has created a sense of virtual community amongst its listeners; the popularisation of the Internet since the early 1990s is encouraging a similar sense of inclusiveness, particularly through fan discussion and networking sites and file-sharing by both musicians and audiences alike. However, if such virtual communities promote a feeling of inclusion for listeners, it is in live performance that music retains most strongly its aura of sacred ritual.

Listening to music at outdoor festivals, particularly in rural surroundings, can enhance this idealised sense of extended community, from Woodstock in 1969 to the rave scene of the 1990s. Simon Frith writes that, 'Under certain circumstances [...] music becomes a source of collective consciousness which promotes group cohesion and social activities that in turn have political consequences' (2000: 319). New Age environmentalism, such as that advocated by Charles Keil and Mickey Hart, takes this notion of collective empowerment a step further, beyond its usual humanism and anthropocentrism, to speculate on a notion of planetary unity that includes care for the natural world, as well as for human beings. As we will see in the next section, the influence of Eastern religions on such thinking, and on the music informed by it, is marked.

## 10.2 Eastern music and environmental consciousness

The idea that Eastern religions such as Chinese Taoism and Zen Buddhism are more ecologically benign than the philosophies of rationalism and scientific materialism that came to dominate Western societies since the time of ancient Greece is a commonplace of deep ecology and New Age thinking. This is despite the fact that, as J. Baird Callicott points out, so-called 'Eastern Wisdom' is made up of many, often competing, philosophies, and that the binary opposition between 'West' and 'East' tends to homogenise the large cultural differences that exist within each geographical area (Callicott and Ames 1989: 13). More importantly, such claims about Eastern ecological wisdom have been questioned by many anthropologists and environmental historians, whose studies of actual social practices show that the high-sounding principles expressed in Eastern religious texts do not necessarily translate into benign practices towards the natural world. For example, China's historical record of environmental destruction predates its recent industrialisation, and suggests, as J. J. Clarke puts it, 'either that the Chinese were not very good at putting their ideals in practice, or even that these ideals are not as benignly disposed towards nature as is sometimes thought' (2000: 87). Anthropologist Martyn W. Lewis similarly observes that the 'Chinese taste for bear paws and gall bladders, monkey brains and snake skins, elephant tusks and cat pelts is entirely indigenous – a fact conveniently ignored by the scores of environmental articles extolling the ecological virtues of Taoism and Buddhism, and informing us that the materialism (or, in some versions, the biblical tradition) [...] of the West lies at the heart of the human assault on the global ecosystem' (1992: 244).

Such considerations tend to be blithely ignored by the advocates of the mystical and idealist versions of ecophilosophy featured in this book. Another such advocate was the German jazz producer and writer Joachim-Ernst Berendt, whose *Nada Brahma: The World Is Sound* (1983) made large claims for music inspired by Eastern religious ideas, by drawing on now familiar conceptual parallels between Eastern religions and quantum physics, which are both considered to view the universe as a form of vibrational energy. Berendt then argued that world music, within which broad category he included not only indigenous non-Western musics, but also jazz, minimalism and rock, represents 'kinds of music that correspond to this sound character of the universe in a most fascinating way' (Berendt 1988: 152).

Significantly, the Foreword to *Nada Brahma* was written by the radical ecophilosopher Fritjof Capra. A cultural shift 'from eye to ear', he wrote, 'would be consistent with the shift from masculine to feminine values that has often been associated with our cultural transformation – from analysis to synthesis, from rational knowledge to intuitive wisdom, from domination and aggression to nonviolence and peace' (xiii). It should be noted that the value judgments Capra makes here, as he privileges the second term in each pairing, in one breath reject not only domination and aggression, but also analysis and rational knowledge. Yet both of these cognitive functions are surely necessary if effective responses to environmental crisis are to be made.

Inspired by Capra's New Age ecologism, Berendt describes two aspects of what he calls the 'new consciousness of music.' Western audiences have discovered the music of Asia, and of India in particular; in addition, the Western world itself 'has come up with a new kind of music that – for the first time in about a thousand years, that is, since the time of Georgian chant – parallels the consciousness of Asian music and yet is completely contemporary and Western, and that neither is nor wants to be influenced by the horrible dictum of Theodor Adorno, that the inhumanity of the arts has to surpass the inhumanity of the world' (152). For Berendt, then, as for Theodore Roszak and Suzi Gablik, it is necessary to move beyond the confrontational negativity of Adorno's modernism if art is to encourage ecological health. Berendt argues that musics based on Eastern modal scales and drones hold the key to such spiritual transformations, because they are less reliant on chord changes and harmonic progressions, and so sound less nervous and more tranquil than most tonal Western musics. Like Susan McClary, therefore, he rejects the notion of teleological structure that for Ernst Bloch was a source of music's utopian potential.

American musicians have indeed taken influences from Indian musics, in styles ranging from classical to jazz and rock, and in doing so have speculated on the relationship between music and environmental awareness. The music of trumpeter Jon Hassell is a synthesis of North Indian influences with indigenous musics from all over the world, which he combines with elements of jazz, funk, minimalism, electronica and hip hop. Hassell speculates on such musical hybridity in terms of what he calls the 'Fourth World', which is an explicitly ecological theory of music.

Hassell's notion of the Fourth World is a form of climatic determinism. Cold-climate tribes, he argues, 'had to develop technology in order to control a hostile environment, and now that very technology has developed in ways that enable them to impose their attitudes on warm-climate tribes (who have, quite naturally, evolved in other important but undervalued ways)' (Hassell 1982: 7). According to Hassell, in the new age of electronic information, technology has developed sufficiently for a new culture that combines the best of both worlds. The Fourth World is therefore a synthesis of the cultures of the cold North and the hot South, brought together as a response to the media-saturated environment of our contemporary, globalised world.

Hassell follows McLuhan in claiming that the processes of globalisation are paradoxically leading to retribalisation. This will solve the crisis of overpopulation when what he calls 'high-density future population' breaks up into 'clusters of New Tribal territories' (7). Hassell gives this process of retribalisation a bioregionalist slant, suggesting that the 'corporate musical imperialism which irons out regionalisms in its drive toward worldwide musical hits in Western pop style' may run into popular opposition that will result in 'a return to a multiplicity of musics arising from tribes of like-minded people once again living within boundaries formed by hills and river beds (like the Tuareg nomads …), linked worldwide by satellites' (4).

The political ramifications of this high tech bioregionalism are that 'we should make an effort to preserve the remaining 'pure' traditions with the same concern shown towards the works of (e.g.) Michelangelo – with an ear towards understanding how these sounds make the day brighter and give courage before the hunt; to understand which music made sorrows bearable and expressed the mystery of creation *before* the entry date of the first transistor radio into the village' (7). It is mainly as a 'pure' cultural resource, then, that Southern cultures will contribute to the Fourth World synthesis.

Before discussing the environmentalist implications of Hassell's theories of music and ecology, let us take a closer look at his music. Having studied raga in the early 1970s with the Indian singer Pandit Pran Nath, he learned, he said, that music is 'shape-making'. Hassell combines this interest in short melodic lines with electronic jazz-funk cross rhythms reminiscent of early seventies Miles Davis, wanting thereby, in his own words, to add to Indian raga a 'kind of earthiness, an urban quality that you find in the African-American approach to things – that synthesis of high and low, that was brought to things by Miles, among others' (Boon nd: 2). In addition, minimalism taught him about 'keeping one thing constant while other things move' (Boon nd: 11). His characteristic trumpet sound is understated, breathy and heavy with reverb, evoking simultaneously human voice and machine: the cyborg fusion that Donna Haraway celebrates, as we will see in the next chapter on 'electronica'. The final layer in Hassell's music is made up of electronic samples of non-Western musics.

Hassell's ongoing Fourth World project began in 1980 with *Possible Musics: Fourth World Vol. 1,* a collaboration with Brian Eno, followed by *Dream Theory in Malaya* (1981). The third album in the series was *Magic Realism* (1983), subtitled 'Aka-Darbari-Java' to denote the range of musical borrowings and influences out of which it is composed: Senegalese drumming recorded in Paris, Indian raga and Javanese gamelan. The music is a dense collage of simple, electronically distorted trumpet motifs, often made up of only two or three notes, repeated over keyboard arpeggios, rhythmical *ostinati* on African drums or xylophone, and sampled 'ethnic' vocal choruses. Heavy reverb defamiliarises the source materials. Each of the instrumental pieces on the album has the nervous stasis of minimalism, gradually building in trance-like intensity while refusing movement towards a linear climax. In this way, then, it fulfils the criteria for the non-hierarchical, anti-Faustian music advocated by Susan McClary and Deleuze and Guattari.

Hassell wrote in the liner notes to *Magic Realism* that 'Aka-Darbari-Java' is 'a proposal for a "coffee-coloured" music of the future – both in terms of the adoption of entirely new modes of structural organisation (as might be suggested by the computer ability to re-arrange, dot-by-dot, a sound or video image) and in terms of the expansion of the "allowable" musical vocabulary in which one may speak this structure – leaving behind the ascetic face which Eurocentric tradition has come to associate with serious expression'.

If these stated intentions are taken at face value, then a gap appears between Hassell's theory of his own musical practice and the music itself, which, in its emotional coolness and affectlessness, surely remains 'Northern' according to his own definition. The trumpet playing has the sparseness and concision of Miles Davis, but lacks the syncopated groove that gave the latter his 'soul'. Hassell has spoken of the way Western music neglects the sensual aspect of music in favour of a concern with abstract structure. In his own music, he wants to combine 'structure with sexiness', and cites Indian classical art as an example, 'with its refined sensuality, in which there is no difference between spiritual and sensual'(Boon nd: 2). The 'equator' between North and South, he told journalist Marcus Boon, 'is the belt line … There's not proper respect for the "gifts of the South", shall we say' (4). Yet in Hassell's hands, world music tends not to be an exuberant dance music; his rhythms lack the groove that animates his original sources, whether African drumming or Javanese gamelan, and is central to works such as Mickey Hart's *Planet Drum*, which we discussed earlier in this chapter.

However, if Hassell's music does not fit easily with his own theorising about it, there is, of course, no reason why it should. Although composers' theoretical speculations may provide a conceptual space within which they create their music, the music itself, like all art, transcends its creators' intentions. Viewed in this way, Hassell's music is a pleasurable and challenging experiment in its own right: its sonic textures and immersive rhythms create not so much a synthesis as a creative clash of traditional non-Western and modern electronic styles, which his collage techniques play off against each other as counterpoint rather than fusion.

From an environmentalist perspective, Hassell's theories of music and ecology are contentious, and raise issues concerning cultural imperialism and Orientalism. Firstly, he has nothing to say about the unequal division of wealth between North and South, which, as the global environmental justice movement points out, seriously impoverishes Southern cities (Athanasiou 1997). Moreover, his idea that Southern cultures should stay 'pure' and free from modern technologies such as the radio, avoids the complex political implications of apparently denying modernisation to the South. His notion of Southern cultures as instinctual and pre-technological also risks racial essentialising and stereotyping.

In Timothy Taylor's terms, Hassell's appropriation of Indian music is another case of the decontextualised appropriation of indigenous musics by a Western composer. In Hindi societies, music is a spiritual discipline, a way of reaching God through what sitar master Ravi Shankar refers to as 'a complete shedding of the ego'. Learning classical Indian music, he observes, requires the disciple to learn 'total humility' in an act of surrender to the guru (Berendt 1988: 156). When American musicians like Hassell have adopted Indian musical styles, however, they have often done so with a lesser degree of adherence to the religious practices that originally informed them, and in some cases no adherence at all.

Hassell is thus open to accusations of Orientalism. According to Edward Said, Orientalism depends for its success on a state of 'flexible *positional* superiority, which puts the Westerner in a whole series of possible relationships with the Orient without ever losing him the relative upper hand' (1985: 7). Indeed, musicologist John Corbett has applied Said's critique of Orientalism to Hassell's music, which he sees as a straightforward example of cultural imperialism. The superficial appeal of Hassell's notion of Fourth World, he writes, 'tends to veil any power politics inherent in such a program, to bury the intricate hegemonic relations between dominant Western musical ideology and local music cultures worldwide' (Corbett 2000: 176). Corbett particularly criticises the way Hassell, like John Cage before him, perpetuates two common Orientalist discourses – the 'wisdom of the East' and 'East meets West' – which serve to mask the reality of what are, in reality, one-sided encounters dominated by Western power interests (179).

Corbett's negative criticism of Hassell may be complemented with John Hutnyk's Marxist analysis of the multi-culturalist ideology of WOMAD, the World of Music and Dance festivals which were founded in Britain by Peter Gabriel in 1982, and which now take place annually throughout the world. Applying Adorno's analysis of commodification, Hutnyk argues that WOMAD's optimistic tone of cultural diversity only masks the ugly, exploitative workings of the global capitalist system. WOMAD, he concludes, may include 'moments of collective subjectivity resistant or unavailable to commercialisation', but 'subsequently these moments suffer the concurrence of entrepreneurialism, industrialisation, bandwagonism, collaboration (and) opportunism' (2000: 47). The festival thus presents a complacently reassuring notion of multi-culturalism that ultimately only helps the further circulation of global capital.

Hutnyk's Marxist rhetoric may be uncompromising, but it offers little in the way of concrete alternatives. The task for a progressive politics, he concludes vaguely, 'is to name the shape and specificity of the current conjuncture and explore and extend ways to intervene' (233). Yet it may be that the entrepreneurs and small businesses behind the success of WOMAD represent attempts at the more socially responsible form of capitalism advocated by environmentalists such as Paul Hawken, even though such a concept would be a contradiction in terms to Marxists such as Hutnyk and Adorno. It remains to be seen whether the world music subculture is in the vanguard of new ways of thinking and acting in a globalised world, which will include 'greener' approaches to the production and consumption of commodities.

Writing of the fantasy version of India invented by the American rock counterculture in the 1960s, Julie Stephens provides a more nuanced view of cultural appropriation than that given by Corbett and Hutnyk. She praises the postmodern use of 'parodic quotation and discordant collage' which characterised the counterculture's subversive, imaginative play with inherited cultural forms. The word 'mantra,' she points out as an example, accrued an open, elastic meaning in hippie culture, so that even rock songs could be included within it: the International Society for Krishna Consciousness held Mantra Rock Dances in Haight-Ashbury in 1967. The counterculture thus appropriated an idea of India as a pristine, pre-modern utopia which represented freedom from enslavement to possessions, while at the same time the notion was paradoxically commodified in records, clothing, and posters. As Stephens observes, 'A remarkable ignorance of both the scriptural and empirical India gave the counterculture the room to fashion a new political language from someone else's terms' (Stephens 1998: 67-8).

Stephens's analysis provides an alternative to simplistic Orientalist critiques which tend to view all Western encounters with non-Western societies as wholly negative and destructive, and all non-Western societies as innocent victims of Western power. Such sweeping,

Manichean notions leave little room for the possibility that cultural encounters between West and East may be equal and mutually enhancing. Historian John M. MacKenzie also gives a more nuanced view of global cultural encounters between the West and its supposed 'Others' than is often given by Orientalist critics. 'At no time', he observes, 'has this repeated appeal to the East been monolithic and seldom has it been reactionary: its phases do not march apace with the rise of western imperialism' (MacKenzie 1995: 170). Instead, he points out that European Romanticism viewed encounters with the Orient as a means towards the regeneration of the West. The Western discourse of Orientalism, he writes, 'seems to go further than merely highlighting the alleged superiorities of Europe. It can modify and therefore surely even challenge the West'. Indeed, such discourses have the capacity to become 'the tool of cultural revolution, a legitimising source of resistance to those who challenge western conventions, introspection and complacency' (10).

MacKenzie's optimistic view of cultural exchange squares with similar attitudes within the world music subculture, according to which, in Simon Frith's words, musical creativity 'always involves cultural borrowing' (Frith 2000: 312). An ongoing problem remains the inequality in financial rewards between white Western pop musicians and their counterparts elsewhere in the world. But in terms of environmental politics, the recently emerging rhetoric of ecological concern in world music, evidenced in its recurrent use of images of a globalised Planet Drum or Whole Earth, does not deserve to be dismissed as mere New Age mysticism. Timothy Taylor concludes that the 'intersecting discourses' in world music of 'authenticity, exploration, emotionality, universalism, spirituality, and others have an immense inertia, which even the most reflexive and sensitive musician would find difficult to overcome and which most of us as listeners succumb to in one way or another' (1997: 30). Another way of putting this, is that there may be something in these 'discourses' after all: they are lived experiences, and not simply linguistic constructions. In its expanded notion of community, which extends beyond the human towards the natural world, world music helps to keep alive the idea that music has, in Ernst Bloch's terms, a surplus meaning which transcends, even if it does not replace, its value in the marketplace.

# Chapter Eleven

# Electronica

'Machinery is aggressive', wrote Ralph Waldo Emerson in 'Works and Days'. 'The weaver becomes a web, the machinist a machine. If you do not use the tools, they use you' (1870: 158). Recent ecophilosophical thinking on music, focusing in particular on the role played by the digital computer in the production of music, has repeated similar Romantic fears that people become dehumanised by the machines they use. This chapter explores the ecophilosophical implications of the use of technology in American electronic music in styles ranging from ambient to rave.

During the heyday of modernism in the early twentieth century, artists working in many fields came to view human creativity as analogous to machine production. Collage in painting and montage in film explored urban modernity as a new perceptual experience of multiple, simultaneous and accelerated space-time. Machine metaphors were not solely the province of fascists like the Italian Futurist F. T. Marinetti, but were also employed by left-liberals such as the American poet William Carlos Williams, who in 1944 defined a poem as a 'small (or large) machine made of words' (1954: 256). Modernist composers as diverse as the Italian Futurist Luigi Russolo, and, in the United States, George Antheil and Edgard Varèse, used machine sounds for their unusual and pleasurable timbres and rhythms (Kahn 1999: 57, 125). The creative role of machines in modern composition was further extended after World War Two, when Frenchman Pierre Schaefer invented *musique concrète* using tape recorders originally developed for military purposes (Griffiths 1995: 17-8; Lange 1998).

In the 1960s, American popular musicians embraced the new, increasingly complex studio technologies, including multi-track tape recorders, to create their own sonic collages with multiple spatial layers. In the autumn of 1967, The Grateful Dead recorded *Anthem of the Sun*, which combined tapes of live shows with studio takes. Guitarist Jerry

Garcia commented that the record was a 'collage' made by 'assembling bits and pieces towards an enhanced nonrealistic representation' (Troy 1994: 117). The record was not a realistic document of a sonic event, then, but a sonic event in itself. As we saw in Chapter 7, rock bands like The Grateful Dead and Jefferson Airplane made explicit connections between the new electronic technologies and an emerging rhetoric of ecological consciousness. The electronic dance music scene of the 1990s revived such thinking, as this chapter will show.

Yet cultural anxieties remain about new technologies, particularly since the invention of computers which have the capacity to play music that is either too fast or too complex for human players. Looked at positively, the pianola music that Conlon Nancarrow composed after 1940, and the rapid drum beats of black British drum 'n' bass in the 1990s, both explore how human creativity can be extended by machines. But for musicians and critics more aligned to Romantic, organicist notions of creativity, the popularisation of the synthesizer since the 1960s has been an unhealthy development, and even an anti-ecological one. Computers certainly are a highly polluting technology: their production releases toxic compounds and heavy metals into the environment, while competitive pressures to upgrade lead to enormous global problems of waste disposal, including political conflict over the unequal distribution of environmental risks (Betts 1994: 28-35). But the idea that the music people make with computers is itself anti-ecological is based on more questionable assumptions.

When New Age pagan Lone Wolf Circles surveyed the current popular music scene in the early 1990s, he found only the 'mundane and meaningless, the stupefying'. It is, he complained, 'an aural paradigm categorising the meaninglessness of contemporary urban life. Each form of music is distorted in the static and mechanical motion of industrial reality.' Widespread use of the synthesizer was the main cause of this aesthetic and ethical decline. Synthesized drum and bass, he argued, have 'robbed most country music of its responsive intimacy and rural flavour, ripped the spontaneity and emotive heart from the breast of New Age jazz'. Jazz itself had degenerated into 'staccato bursts echoing the nearby jackhammers, and speeding cars' passage' (Lone Wolf Circles 1991: 54).

Aesthetically, Lone Wolf Circles may have a point about the crude use of synthesizers in some contemporary popular music. As Evan Eisenberg notes, to describe music as mechanical implies 'rigidity and insensitivity to nuance', a charge that can convincingly be made against some of the New Country and New Age jazz of the 1990s (1987: 145). But Lone Wolf Circles' aesthetic dislike of synthesizer music also seems to be informed by a deeper, political opposition to urbanism and modernity in general. That music might echo the rhythms of jackhammers and cars seems, for him, to be wrong in principle. Instead, he quotes approvingly the view of the Greek pianist and composer Sakis Papadimitriou, that acoustic instruments are more ecologically healthy than electronic ones:

> Today, playing the flute, congas, piano, saxophone, becomes an ecological protest. Sound emerges from within the musical instruments, directly from the source. And you hold this source in your hands, in your lap, in your arms, around your neck. Vibrations warm you. The sound touches you. It is not coming at you from a different, distant and mechanical point. Synthesizers do not emit music. They are faulty. They do not vibrate. The sound takes a digital walk first and them jumps into the loudspeakers. This is the only way they come to linger in the air wearing their spacesuits. These sounds are ignorant of the earthly environment. They do not possess the elementary quality of surviving in nature (Lone Wolf Circles 1991: 56).

For Papadimitriou, music should be an extension of the body of the musician; instruments should produce sonic vibrations out of the resonance of the physical materials from which they are made. By denying such physical origins, music made on computers is not only unnatural, but is perpetuating a dangerous attitude that threatens human survival itself. From this perspective, using the synthesizer is wrong both aesthetically and ethically.

However, there are more positive approaches to the role of the digital computer in music, particularly since the popularisation of the personal computer in the 1980s began to radically transform the democratic potential of computer technology. For the British musician and writer David Toop, the new 'virtual' music created on PCs and laptops creatively explores what it is like to live in the contemporary electronic age. The wide range of creative actions within digital music culture, from sampling and glitching to the use of the Internet for exchanging sound files and facilitating shifting collaborations between musicians, are rhizomatic activities, in Deleuze and Guattari's sense, with the potential to challenge rigid social hierarchies rather than simply reinforce them (Toop 2004: 198).

While Sakis Papadimitriou's concern, endorsed by Lone Wolf Circles, about the decline of physical disembodiment in music, is important, as Toop notes, older electronic instruments such as the electric guitar also enabled music to be made without the need for a natural resonance chamber. Yet the digital computer takes this process of disembodiment a step further, so that the physical embodiment and location of the performer is less important for the music's aesthetic value than what Toop calls the 'consciousness of the musician' (232).

Paul Levinson further emphasises the novelty of recent technological developments. The digital computer can be distinguished from earlier electronic instruments because of the way it establishes a new relationship with the body of the musician. He observes that the electrification of the guitar did not put an end to the need for human touch when playing the instrument, and therefore to the virtuosity that links the Spanish acoustic maestro Andres Segovia with Eric Clapton. But digital music made on a computer has eliminated the need for touch as a physical determinant of the music.

Levinson sees this new situation as culturally liberating, because it concentrates critical attention on the quality of the musical idea itself, what he calls 'the mental creation of the music – the hearing of it in the performer's head,' rather than on its execution. Computerisation, in this light, has the potential to democratise the means of musical creativity by making physical technique less important than the creative idea behind it. Indeed, Levinson goes on to extend the significance of such technological developments to the culture as a whole: 'This liberation of the mind from its immediate dependence on tangible things for expression of its creativity is indeed the key to how the digital revolution may be making everything better, and not just in music' (Levinson 1999: 169).

Compared to such eulogies for the bright digital future, Toop is more ambivalent about the loss of the haptic and the organic in the creation of music. While largely embracing the creative potential of digital music, he also values the physical intimacies of improvised performance. He is thus drawn to the open-minded, inclusive view expressed by the Icelandic singer and composer Björk, that 'you can be organic and pagan and have ProTools' (Toop 1995: 226). Indeed, that electronic composers are as fascinated with the natural world as they are with their computer software programs is a theme we will explore in the rest of this chapter.

## 11.1 Popular music and environmental sound

In 1968, producer and songwriter Van Dyke Parks suggested to Bernie Krause and Paul Beaver that ecology would make a good theme for a synthesizer album, and encouraged them to use field recordings as compositional elements. In that way, as Krause puts it, 'we could use our art to address the need to preserve habitats and use the sounds as a new source of music for our craft' (Krause 1998: 56). Krause described the resulting album, *In A Wild Sanctuary* (1970), as 'environmental impressions', and claims that it marked 'the first time unaltered natural sound had been directly used as a component of orchestration' (59).

Environmental sounds are featured on two tracks. On 'So Long As the Waters Flow', the sounds of crashing waves punctuate the progressive rock structure. 'Walking Green Algae Blues' includes field recordings of streams, birds, people and animals at the San Francisco zoo in a slow, bluesy call and response between Hammond organ and electric guitar. A commentary at the zoo can be heard announcing that orangutans 'are now rare in the wild because of persecution by man'. Given such connotations, the clanging of a San Francisco street car, faded into short-wave radio interference, sounds sinister and unnatural. A sudden thunderclap confirms the mood of apocalyptic endangerment. Yet, as in some of the similarly apocalyptic rock songs written at this time, which were discussed in Chapter 7, the overall effect is of bathos rather than dread.

David Rothenberg comments on the use of taped nature sounds in music: 'If we use nature, we must really listen to what we use and put it to use in a way that respects its own life and integrity – if that is at all possible' (2001: 8). That Krause admits to using the sound of a flushing lavatory on 'So Long As the Waters Flow', because it made a better water sound than his field recordings, does not suggest a concern with authenticity that would please Rothenberg (Krause 1998: 57). Nor do Krause and Beaver allow their field recordings to alter the received blues-based rock format on which the piece is based. As a result, the environmental sounds function decoratively rather than structurally. The sound of crashing waves is not integrated into the rock framework, but rather is dominated by it. Despite its obvious good intentions, then, *In A Wild Sanctuary* comes across as both kitsch and bathetic in its use of nature sounds.

In contrast, John Hudak and Paul Lansky transform the sounds of everyday spaces into new and unfamiliar textures; the cold, metallic timbres of their electroacoustic soundscapes can be delicate and fragile as well as uncanny and disturbing. For 'Night Traffic', on Lansky's 1993 album *Homebrew*, the composer used a synthesizer to impose musical harmonies and timbres onto the sound of traffic on his local four-lane highway. The resulting deep drones and trebly arpeggios suggest an electric organ or piano, and give the source sounds an ethereal quality as they build with the harmonic drama of progressive rock.

Hudak's music draws on sounds from both rural and urban sources. His technique is to manipulate a single sound source in order to explore the essence of that sound. 'Sketch of a Field' (1997) slows down and distorts the sound of crickets into the roar of white noise. As the security of referentiality is removed, a disturbing, dream-like mood is established which reveals an uncanny dimension in everyday sounds. 'Brooklyn Bridge' (1998) uses contact microphones to capture the resonant hum of the bridge itself, as if it was being played by the traffic moving over it. 'Highway' (2001) uses a very low mastering volume to turn traffic noise into shimmering, treble sonic patterns and sinister bass rumbles.

Hudak compares his electroacoustic compositions to Japanese haiku, which he considers 'a record of a moment keenly perceived which in some way links Nature with human nature'. 'I enjoy looking into the smaller sounds, the ones most people can't hear, don't hear, or don't listen to', he told an interviewer. 'So my interest is in clarifying, filtering, framing these sounds to make them audible'. Hudak's lack of interest in the social meanings of the sounds he uses recalls John Cage (Ingram 2007a). 'My choosing the Brooklyn Bridge and the Brooklyn-Queens Expressway (for 'Highway') as sound source material', he commented, 'has nothing to do with their place in automotive culture. Rather, I am exploring the sound environment of my immediate world: the Brooklyn Bridge is a defining symbol of New York City and is visible from my window, while the Brooklyn-Queens Expressway runs underneath my apartment.' 'My aesthetic aim', he added, 'is devoid of any socio-political overtones or constraints' (Haynes 2001: 14).

That said, Hudak's work still generates meanings that transcend his own intentions. In their electro-acoustic transformations of traffic, both he and Paul Lansky invite the listener to experience such everyday, urban phenomena in a fresh way. Yet making beauty out of the sounds of urban environments, including those of automotive culture, has social and political implications. Indeed, such implications are what make aural celebrations of modernity unacceptable to Lone Wolf Circles, who, as we saw earlier, does not want to listen to music that sounds like cars or jackhammers.

The electroacoustic music of Lansky and Hudak employs modernist techniques of defamiliarisation to challenge their listeners' habitual perceptions of the sounds of their immediate environment. In the 1970s, the genre of ambient electronica emerged in popular music to explore what music therapist June Boyce-Tillman calls a more 'nurturing' mode of listening. It is to these developments that we will now turn.

## 11.2 Ambient electronica and the simulation of nature

Cox and Warner define 'electronica' as an 'unlikely meeting' of 'the sonic and intellectual concerns of classic electronic music; the do-it-yourself and bruitist attitudes of punk and industrial music; and beat-driven dance floor sounds from disco through House and Techno' (2004: 365). 'Ambient electronica' is itself a sub-category of 'ambient music', a term first coined by the British composer Brian Eno in the early 1970s to refer to the music he chose to listen to 'for its stillness, homogeneity, lack of surprises and, most of all, lack of variety' (1996: 293). Ambient music provides the immersive listening experience whose ecological dimensions were discussed in Chapter 4; as Eno puts it, it is music 'to swim in, to float in, to get lost inside'. It is intended 'to induce calm and a space to think' (294). As such, ambient music has antecedents in the multi-million dollar Muzak industry, which, as Joseph Lanza has shown, provides 'environmental music' intended to work subliminally on its listeners to make manual tasks easier to perform (2004: 120). The sonic simulation of natural landscapes in ambient music also recalls the 'exotica' of Les Baxter and Martin Denny, whose lushly orchestrated instrumentals evoked Orientalist tourist paradises for suburban American consumers in the 1950s (Hollings 1998: 40-5).

Ambient electronica also overlaps with New Age music, a form that has few defenders amongst music critics. An exception is June Boyce-Tillman, who values its therapeutic function. 'If we accept that music can do this', she writes, then music can become 'a backdrop to other internal action within ourselves', rather than 'something to be concentrated on'. New Age music thus 'provides a security in which new ideas can be explored and new risks taken' (Boyce-Tillman 2000: 54-5). Boyce-Tillman's view is refreshing in its lack of musical snobbery, yet it is a minority view, particularly among critics on the Left. Marxist geographer Neil Smith refers disparagingly to the compact discs and videos sold by the Nature Company as the 'genre of nature as psychic soporific' (1996: 37). For him, the electronic pastorals evoked by New Age music are an opiate rather than a useful therapeutic. Music critic Simon Reynolds writes that New Age music, by privileging melody over rhythm, or the head over the body, is a repressive, 'white' form of music (1999: 182). Yet this attitude fails to recognise both the role of melody in many forms of 'non-white' music, and the crucial interest in bodily health in the music of New Age composer Steven Halpern, for example. Nor is the melodic or harmonic simplicity of New Age music a problem in itself. The real point is that some New Age music can be too predictable, melodically and harmonically, and therefore lacking in aesthetic interest, whatever therapeutic functions the listener may use it for. On the other hand, Halpern's *Chakra Suite* (2001) achieves aesthetic interest in both the sensual pleasures of its sound textures, for electric piano, synthesizer and Tibetan Singing Bowls, its meditative play with duration, and its use of minimalist melodic and harmonic techniques. Authentic New Age music, Halpern writes, 'employs time, space, and silence as a sonic vehicle to get the listener into closer contact with his/her spiritual nature' (Halpern 1998: viii). Halpern's Pythagorean interest in the healing potential of music, its potential for what Enrico Fubini calls 'allopathic catharsis', is an important contribution to the postmodern eco-aesthetics we have seen advocated by Suzi Gablik, and extended into music criticism by June Boyce-Tillman, the potential of which for ecological practice should not be underestimated.

For Lone Wolf Circles, ambient electronica tends to be a poor simulation of the real natural environments to which, as program music, it relates. Its cultural role, according to the New Age neo-pagan, is to compensate human beings for the loss of the outdoor existence which our ancestors apparently enjoyed, and to which, he asserts, we truly belong. He thus accuses ambient music of being a form of 'minstrelsy', an offensively weak and inauthentic imitation of the real thing. Whereas the best ambient music works, he writes, as 'wordless, adagio, the milieu of receptivity moving slowly like a wide jungle river – time to see everything that passes, unhurried, without comment', too often 'the composer fails as a catalyst of perception, ministering to a kind of New Age Muzak, wrung out of the latest in phonic synthesizers' (Lone Wolf Circles 1991: 56). We are back with the author's apparent phobia for synthesized music in itself with which this chapter began.

An early example of ambient electronica which avoids the predictability of New Age Muzak is Wendy Carlos' *Sonic Seasonings* (1972), in which the composer explored the musical potential of the new Moog synthesizer. The record, wrote Rachel Elkind in the original liner notes, was 'designed to be part of the decor, so to speak – a sonic ambience that enhances the listener's total environment'. What the listener needs to bring to the record, she concludes is 'your own imagination and his remembrance of Nature's blessings' (Elkind 1998: np). Electronic pioneering thus encourages nostalgia for unspoiled Nature. As program music, the slowly changing, immersive sound of the four pieces of music encourages both the contemplation of imaginative landscapes and a fascination with the nuances of sound itself, including the slippage between electronically synthesized sounds and real sounds, such as rain, thunder and bird calls, used as *musique concrète*. Carlos's evocations of the natural world tend to avoid the prettiness and securities of the pastoral. Instead, 'Summer' suggests the oppressive heat of midday through slow, claustrophobic arpeggios, whereas in 'Winter', Nature is inhospitable to human beings: field recordings of wolves howling punctuate complex synthesizer layers of slowly moving modal scales, augmented chords and beatless rhythms. This is ambient music that is complex and tonally unsettling.

Even though it was released in the early 1970s, when, as we saw in Chapters 6 and 7, American folk and rock musicians were bringing a sense of ecological endangerment into their work, *Sonic Seasonings* makes no reference, in either its supporting texts or the music itself, to ecological or environmental concerns. In this absence, it may be contrasted with Steve Roach's *Desert Solitaire* (1989), which is another example of ambient electronica that, like Carlos' work, is sensitive to musical nuance, and provides the space for contemplative listening in a way valued by Lone Wolf Circles, June Boyce-Tillman and Brian Eno. Gradually shifting minor chords and sound washes suggest the sublime grandeur and melancholy of the deserts of the American west; slow *ostinati* evoke a sense of the uncanny rather than of repose. The title's reference to writer and environmental activist Edward Abbey also suggests that not all New Age ambient music is politically quietist or conservative in the way suggested by Neil Smith. Rather, a path evidently exists out of the simulated aural environment to the politics that shape the real landscape outside the recording studio.

In a night club setting, ambient electronica is a relaxing, 'chill out' music that provides a space for rest and conversation. In this way, it complements the loud, repetitive beats of electronic dance music, which, as we shall see in the final section of this chapter, also had advocates in the rave subculture of the 1990s as a means of reconnecting its audience with the natural world.

## 11.3 Organicism and electronic dance music

In the early 1990s, the British record industry began to use the term 'world dance fusion' to distinguish acts such as Transglobal Underground and Loop Guru, who mixed identifiably 'ethnic' samples and instruments with a strong, disco-derived electronic dance beat, from 'world music', whose more purist folk styles were popular with an older, more middle-class audience, but had never caught on with the younger, working class audience for electronic dance music. As reviewer Tony Marcus dismissively put it in May 1994, the new 'Ethno Trance' movement offered 'global sounds a second chance after the half-hearted World Music hype of the mid-80s [...] The endless CDs of Eskimo, Tibetan and Asian virtuosi bypassed both the pop and underground music scenes to lurk in a ghetto of right-on vegans and Sunday supplement readers' (Hesmondhalgh 2000: 284). In the 1990s, the electronic dance music scene favoured by Marcus came to adopt the countercultural rhetoric of New Age environmentalism, as we will now see.

In the United States, rave culture did not become a mass youth movement as it did in Europe; nor has the American scene produced a distinctly American version of the music. Instead, the American record industry preferred to promote grunge, which, with its guitar riffs, iconic stars and familiar images of teenage rebelliousness, was a style of music it understood much better. Moreover, many American rock fans considered rave music inauthentic compared to indie-rock and hip hop, and in so doing revived some of the rhetoric of the 'disco sucks' campaign of the 1970s (Reynolds 1999: 307).

Where electronic dance music did catch on in America, however, it attracted an aura of ecological speculation, focused not on the supposed authenticity of traditional acoustic instruments, as in world music, but on the mystical capacities of the digital synthesizer. In the early 1990s, a section of the British acid house scene began to revive the techno-paganism of the sixties rock counterculture and to update its rhetoric of ecological involvement. The American scene revolved around British expatriates such as Genesis P. Orridge and Fraser Clark, who became self-appointed authorities in the New Age theories of music as cosmic vibration we explored in Chapter 4. Unsurprisingly, San Francisco became the centre for such mystical speculations. The rave scene built on the West Coast 'neuroconsciousness' and designer drugs movement, led by Timothy Leary and Terence McKenna, as well as the local tradition of psychedelic rock in Haight-Ashbury and the Californian New Age culture which followed it. What Simon Reynolds summarises as the 'Internet/Virtual Reality/posthuman/'extropian'/*Mondo 2000* scene on the fringes of Silicon Valley' also provided an existing culture for its growth (Reynolds 1999: 314; Rucker et.al 1992; Rushkoff 1994).

Terence McKenna continued Timothy Leary's project of using psychedelic drugs apparently to reconnect human beings with wild nature, both inside and outside the self. His speculations on cosmology and anthropology combined elements of deep ecology and ecofeminism in a grand, cosmic theory of evolution. According to McKenna, human consciousness may have evolved as a result of eating magic mushrooms. Early human beings once lived in a paradisal state based on eating the psychotropic drug psilocybin, ingestion of which led to a sense of 'boundary dissolution' that sustained an anarcho-utopian tribal society, organised around orgiastic mushroom-eating ceremonies enacted every full moon. Eventually, climatic changes led to the demise of these primal mushroom cults, and a fall from Edenic innocence.

Drawing on ecofeminist Rianne Eisler, McKenna argued that the subsequent emergence of the ego, which had been inhibited in primal society by the effects of psilocybin, brought with it a new 'dominator' form of social organisation based on hierarchy, patriarchy and materialism (1992: 149). Plant-based hallucinogens were thus a cure for the pathological ego on which the modern era is based. For McKenna, who died in 2000, popular music could play a role in the 'Archaic Revival'. Rave music was a trance-inducing cult which could end our alienation from what he called the 'Gaian collectivity of organic life'

(149). He told David Toop that electronic instruments 'are being used to invoke a nature that has almost slipped from our cognition. We don't live in that kind of nature. To us, nature is a park. Real nature is very much like the ambience of dense electronic music' (Toop 1995: 278).

A second figure influential in the San Francisco rave scene in the early 1990s was Genesis P. Orridge, leader of the band Psychic TV and its attendant cult, the Temple ov Psychick Youth. Simon Reynolds usefully sums up his ideas as 'psychedelia/sampladelia as the creative abuse of technology; house's 125 bpm as the primordial trance-inducing, alpha-wave-triggering tempo that connects Arab, Indian, and aboriginal music; the manipulation of sonic frequencies to achieve "metabolic engineering", *à la* Aleister Crowley's dictum "our method is science, our aim is religion"' (1999: 150). Interviewed in Jon Reiss' documentary *Better Living Through Circuitry* (1999), Orridge speaks of how rave music has 'reduced down in the West, for the first time, what is so-called popular music and youth culture to a ritual which admits to, and utilises, the most arcane and ancient methods for achievement of altered states, and a celebration of that contact with otherness'.

Most of the electronic music directly inspired by Orridge and McKenna has been European in origin. The Dutch techno group The Psychick Warriors ov Gaia referenced the former, while the British rave group The Shamen's 'Re-Evolution' sampled a speech by McKenna on shamanism and the Archaic Revival. American interest in the ecological implications of rave music has been led by Brian Transeau. As BT, he produced a style of what he called 'epic' or 'dream house' music which was more popular in Britain than in the United States. Speaking of his interest in making 'organic electronic music', he explained that every sound on his records 'is generated by something organic, whether it's my hands, somebody else's hands, a mouth, or the cicadas in the woods [...] it's generated by something natural'. By grounding his music in the natural world, his intention was to move his audience into the trance states experienced by indigenous shamen. Exposure to continual auditory and visual stimuli of a very small frequency range, he argued, leads people 'to entrain to that particular brain state that they're being stimulated with, either by photic flashing or by auditory sound'. In Amazonian shamanic rituals, the shaman beats a drum at 4 to 6 wave cycles per second, thereby entraining the participants to the 'theta' brain state. In club culture, BT claimed, the photic and auditory 'driving' produced by the stroboscopic lighting and repetitive music entrains people into the same creative brain states. In live performance, BT wore a brain-scanning device to monitor his own brainwave activity, so that the music he produced could correspond to these brain states.

'Firewater', the opening track of BT's second album *ESCM* (1997), illustrates well his notion of 'organic electronic' music. It begins with samples of frogs, birds and cicadas, before a deep, spoken voice announces 'a fire in the sky'. A lyrical piano motif is joined by ecstatic, wordless, gospel-inflected vocals, before a shuffling, mid-tempo beat begins. The beat accelerates, stops and restarts, adding an unpredictable edge to the simple tonal harmonies; the mechanical, phased sound of a helicopter interrupts the flow, and the track ends surprisingly with a skirl of bagpipes and strumming on an acoustic guitar. The closing lyric contemplates being 'six feet under', but the song's references to death and apocalypse do not contradict its optimistic New Age mood of transcendence and spiritual uplift.

As we saw in Chapter 4, the notion of rhythmical entrainment on which BT bases his concept of music is central to New Age beliefs in the potential of music to invoke a mystical sense of communion with all living beings. BT's music signifies this feeling of cosmic connection in the title 'Lullaby to Gaia', in which a fast, techno beat and the familiar tension-release rhythms of house music are complemented by sweet female vocals, intoning New Age clichés ('the answer lies within', 'you know the truth'). The expansiveness of the slowly building rhythms and the eclecticism of the sound materials put artistic life into such familiar utopian yearnings.

A leaflet entitled 'House Music and Planetary Healing', circulated by Malachy O'Brien, another British expatriate in San Francisco, amplifies the ecophilosophical context behind BT's linking of house music with Gaian mysticism. 'When used with positive intention', it reads, 'Group energy has the potential to help restore the plan of Love on Earth [...] In the true sense of rhythmic movement, the effect is to align the physical, mental and emotional bodies with the Oneness of All that Is' (Reynolds 1999: 154). House music thus promoted a mystical notion of rhythmical entrainment with the Earth conceived as a benign Gaia. In the American rave scene, outdoor events such as the Full Moon raves best expressed this yearning for a return to a natural arcadia (St. John 2001: 5).

Digital music-making may invite such mystical thinking because digitalisation is an invisible process that manipulates relationships rather than things. David Toop questions the authenticity of such acid house 'technopaganism'. 'I have serious doubts', he writes, 'about automatic assumptions that the kind of free-floating ecstatic states induced by drugs, dancing and probable dehydration are identical to shamanism's cosmologically centred spirit journeys' (Toop 1995: 278).

However, as anthropologist Gilbert Rouget notes, the physiological effects of starvation and hallucinogenic plants are central to the mental changes in 'real' shamans, so that a clear distinction between authentic and inauthentic experiences is difficult to maintain. Rouget usefully distinguishes between 'possession', which is an involuntary state, and 'trance', which is voluntary. In trance events, then, music acts 'as a psychological sign rather than a physiological cause'. He notes that both Beatlemania and religious possession cults fall into the latter category, because in neither 'do subjects fall into trance who are not disposed to do so' (Rouget 1985: 37). The trance effects of popular music are thus not different in kind from those of religious experience, then, although they usually have different meanings projected onto them. As Erik Davis noted of Ken Kesey's experiments with shamanism, he 'generally failed to construct anything like the *contexts of meaning* that traditional shamanic or religious cultures have always used to integrate cognitive ecstasy (and its metaphysical morning-afters) into ordinary life' (1998: 149).

The proponents of technoshamanism, in any case, do not claim to be interested in authenticity; rather, their investigations of the imaginative potential of the new technologies of electronics and drugs to create a sense of global, electronic interconnectedness question inherited notions of authenticity. As anthropologist Scott R. Hutson observes, 'By spinning "tribal" beats on sophisticated equipment, the DJ synthesises our desire to be spiritual with out rootedness and dependence on the material. The DJ thus serves as a model of the place of machines in the world and a model for how the soul can be integrated with them' (Hutson 1999: 71)

However, although partying at a rave event may help communal bonding, it is also an act of consumption. Some members of the eco-anarchist subculture which attached itself to rave music in the 1990s, under the influence of Terence McKenna and Genesis P. Orridge, named their movement 'New Edge', to distinguish it from the less militant philosophy of the New Age, even though it shared many ideas with the latter, such as Gaia and the supposed spiritual dimension of quantum physics. Atomjack, curator of the Fusionanomaly website, wrote that New Edge 'filters out the fluffy flowers and cheesy peacenik offerings and gets down to business. Well, the business of partying anyway ...' (Fusion Anomaly: nd). The latter remark shows the limitations of focusing on popular music alone as a means of social and political transformation. In Ernst Bloch's terms, the ecological mysticism of rave culture expresses 'abstract' rather than 'concrete' utopian yearnings.

David Rothenberg observes that fears of what he calls 'monolithic megamechanism' hinder a better understanding of technology. 'We get little sense of how to deal with the implications of any individual machine', he writes, 'if we simply make it part of a great, frightening conspiracy slipping out of our grasp, engulfing us, impossible to escape or turn off' (1993: xiv). A better notion is 'appropriate' technology, therefore, 'where appropriateness is defined in terms of how well the whole of humanity can extend itself into the environment without destroying it or losing the ability to see beyond our own narrow interests' (xvii). Donna Haraway's trope of the cyborg, those 'fabricated hybrids of machine and organism', is useful in this context for the way it goes beyond the limitations of Romantic naturalism. The cyborg, she writes, 'skips the step of original unity, of identification with nature in the Western sense' (Haraway 1991: 151). As noted in Chapter 3, there is a need for a cultural recognition of both organicist and mechanistic models of knowledge, in order to question the negative connotations of the machine evident in some radical environmentalist thinking.

Despite the reservations with their philosophical approach outlined in Chapter 1, Deleuze and Guattari's theory of the subversive power of the rhizome also has the advantage of combining organicist with mechanistic models. This version of postmodernism, articulated by both Donna Haraway and Deleuze and Guattari, is preferable to the cruder definition of the difference between modernism and postmodernism given by Frederic Jameson: 'In modernism [...] some residual zones of "nature" and "being", of the old, the older, the archaic, still subsists; culture can still do something to that nature and work at transforming that "referent". Postmodernism is what you have when the modernisation process is complete and nature is gone for good. It is a more fully human world than the older one, but one in which "culture" has become a veritable "second nature."' (Jameson 1991: ix).

A musical interpretation of Jameson's postmodernist vision of nature 'gone for good' may be *Second Nature* (1995), a collaboration recorded in Brooklyn, NY between Atom Heart (the German DJ and producer Uwe Schmidt), the Japanese producer Tetsu Inoue and American bassist Bill Laswell. The titles of two of the four tracks indicate the postmodern sensibility: 'Synthetic Forest' and 'Artificial Seaside'. Both tracks play with referential listening: synthesized versions of the sounds of birds, the sea, wind and water punctuate a static soundscape in which melody and beat are largely absent. Several minutes into both tracks, Laswell begins a driving bass riff, and there is some syncopated interplay between the synthesized effects. As the harsh timbres interact, the overall mood is cold, sinister and ghostly. However, though *Second Nature* is not a hospitable soundworld for the listener to inhabit, it is at times fascinating and exhilarating.

If *Second Nature* explores what a totally synthetic world may feel like, it does not suggest profound anxieties about the death of nature. As such, the album differs from the conservationist concerns that underpinned Steve Roach's *Desert Solitaire* or the New Age desire to liberate the natural body in Steve Halpern's *Chakra Suite*. Andrew Murphie has written that the 'confusion of mechanics and biology' in the thought of Deleuze and Guattari is 'often celebrated, and more than that, enactively embodied in electronicas'. Such diverse musics, he argues, explore uses for modern technology that are non-militaristic and collective, and in so doing act as 'therapeutic analysts for the network society' (Murphie 2003: 8). From this perspective, the electronic pieces discussed in this chapter, from *Desert Solitaire* to *Second Nature*, are exemplary actions that, in their different ways, help listeners understand, at the imaginative and emotional level of art, newly emerging relationships between technology, the human body and the natural world.

# Chapter Twelve

# Jazz

Jazz combines composition with improvisation in varying degrees, but it is its emphasis on improvisation, both individual and collective, that has come to enact, for musicians and audiences alike, humanist ideals of democracy, equality and freedom, summed up in the metaphor of the 'collective conversation' (Berliner 1994: 348). For African-Americans, these ideals are inseparable from notions of racial advancement and political resistance (Lewis 2004: 275). In Ernst Bloch's terms, improvisation in jazz represents a concrete utopia to which they aspire, and for which they are struggling. Unsurprisingly, such concerns, when considered from an environmentalist perspective, have been largely anthropocentric and pre-ecological, given the dire need to address issues of poverty and inequality of opportunity for African-Americans in the United States. Nevertheless, in recent years some jazz musicians and critics, both black and white, have begun to interpret their art in terms of notions of ecology, as we will see in this chapter.

## 12.1 Jazz and ecological awareness

Gary McFarland's big band jazz suite *America the Beautiful, An Account of Its Disappearance* (1968) was an early response by a jazz musician to the ecological crisis that came to public attention in the 1960s. The album's liner notes included a passage by Marya Mannes, written in 1958, that attacked Americans who were conserving 'their convenience at the expense of their heritage, and whose ephemeral prosperity was built on waste'. In '80 Miles An Hour Through Beer-Can Country', discordant harmonies and car horns evoke the despoliation of the land. The following track, 'Suburbia: Two Poodles and a Plastic Jesus', is a minor key lament for a lost paradise. Yet McFarland's witty arrangements and funky rhythms sound fun rather than apocalyptic, or rather they make environmental apocalypse sound fun. As we saw in Chapter 7 on sixties rock, the catchy tonal harmonies and entraining rhythms of popular music can have a therapeutic function, which in this context treats anxieties about environmental apocalypse in a reassuring way.

That McFarland's ecological concerns focused on issues of over-development and suburban sprawl reflects the emergence of environ-mentalism in this period as a largely white, middle-class 'quality of life' issue focusing on threats to the American wild or rural landscape. In contrast, African-American jazz has typically celebrated the urban rather than the rural, because the city represented hopes for freedom from the exploitation and cruelty black people had experienced in the countryside under slavery. As literary ecocritic Michael Bennett observes, African-Americans have often 'constructed the rural-natural as a realm to be feared for specific reasons and the urban-social as a domain of hope' (Bennett 2001: 198). Unsurprisingly, then, many African-American musicians have celebrated urban modernity. In 1971, Duke Ellington's *New Orleans Suite* returned to the urban origins of jazz; he described 'Bourbon Street Jingling Jollies' as 'a rhythmic tone parallel to the excruciating ecstasies one finds oneself suspended in when one is in the throes of the jingling rhythmic jollies of Bourbon Street'. The piece is led by Norris Turney's gently lyrical flute which evokes the city as 'pastel enchantment'.

For Sun Ra, the future utopia prefigured by jazz was a futuristic urban scenario inspired by science fiction. As one of his most famous pieces put it: 'Space is the Place'. Graham Lock explains that Sun Ra projected this utopian future 'not as the location of a promised afterlife but, in the light of astronomy and modern technology, as the site of a real, living future in outer space' (1999: 34). Yet space travel was also a metaphor for transcending present suffering through the power of the imagination, as it was for his rock music contemporaries such as The Grateful Dead, as we saw in Chapter 7.

Sun Ra combined his interest in high tech futurism with Romantic appeals to the beneficence of pre-modern people of colour. His philosophy of 'Astro Black Mythology' revealed an interest in theo-sophy in which music was a hermetic mystery that could realign people with the spiritual dimensions of reality beyond surface appearances. It also embraced the religion of ancient Egypt rather than the Christianity imposed on the African slaves in the United States (Szwed 2000: 108-9). Sun Ra's composition 'Ancient Ethiopia', on 1958's *Jazz in*

*Silhouette*, evokes this pre-modern Afro-pastoral through a stately modal theme on baritone saxophone set against Latin-American percussion and a pair of ethereal flutes. The relaxed groove and formal open-endedness provides the potent sense of space central to Sun Ra's music.

Running parallel to jazz's embrace of urban modernity is the presence of the pastoral tradition in African-American music which we described in relation to hip hop in Chapter 9. This construction of the pastoral as a promise of both spiritual and physical transcendence was also characteristic of the 'free jazz' movement of the early 1960s, in which the apparent anarchy of improvised performance came to be seen as a utopian prefiguration of what Alfred Willener called 'future non-oppression, of the absence of an imposed order' (1970: 260). The music of late period John Coltrane, Pharoah Sanders and Albert Ayler, amongst others, conceived of this freedom in spiritual terms, often evoking the ecstatic sense of Oneness spoken of in Hinduism. Coltrane's soprano saxophone introduced Oriental tonalities into jazz, as he became an innovator in what would later be called 'world' music.

Like the jazz-bohemian style of hip hop described in Chapter 9, African-American free jazz has expressed an Afrocentric sensibility which revisions Africa as an unfallen pastoral utopia. The title track of Coltrane's *Kulu Sé Mama* (1966) invoked this notion of Africa through a drifting, open-ended mesh of percussion, vocal chanting and Coltrane's modal blowing. The liner notes featured percussionist Juno Lewis's poem 'Juno Sé Mama', which called for the rebuilding of African-American art and culture after slavery through a reconnection with the natural world. The ritual begins with 'a / Mighty cloud burst / And the rippling of the water drum begins beating against the air cups of the world. Moon children … ready to be born. Signs of sky, earth, water' (Szwed 2000: 108-9).

Like Coltrane's late music, Pharoah Sanders' album *Karma* (1969) also brings to mind both Afro-pastoral and the cosmic sublime. 'The Creator Has A Master Plan' builds slowly into dense, immersive textures, as Sanders' tenor saxophone, at times exuberant, raucous and gentle, converses with the piano and percussion groove. The ensemble-playing creates a sense of spiritual mystery through simple repetitions within a gradually shifting whole. The 'Master Plan' referred to in the title is 'peace and happiness for every man', the refrain sung by Leon Thomas. After this half-hour of ecstatic rhapsody, 'Colours' briefly closes the album with a sentimental and child-like evocation of the African pastoral: 'Mother Nature seems to love us so / When she smiles there is a subtle glow'.

The pastoral utopianism of sixties African-American free jazz was given an explicitly ecological direction by Don Cherry. Speaking in Norway in June 1971, he described to Arthur Taylor the role that he believed the natural world played in his own practice of jazz improvisation: 'People who live in nature and with nature in Africa, China, or in the woods give their music a quality which is both earthy and godly, and that's what I think should be preserved: this quality of earthiness' (Taylor 1982: 176). By adopting musical instruments and ideas from Africa and Asia, Cherry achieved this sense of 'earthiness' in his own music. He drew on amateur folk musics which, he argued, exist outside the false professional and commercial emphasis on individual style that dominates Western music. On his album *Orient*, recorded in Paris in 1971, the scratchy tone of the pocket trumpet he had bought in Pakistan is both lyrical and ecstatic; Turkish drummer-percussionist Okay Temiz lays down complex metres borrowed from Indian music, over which are layered flutes and vocals influenced by Islamic microtonalities.

Asked why he had left the United States for Europe, Cherry replied: 'We live on Planet Earth, and I'm at a period in my life where I have to come close to Mother Earth and live in the rotation of the seasons. I had lived in the cities most of my life, and I reached a point where I had a polluted brain, a polluted soul. The only cure for me was nature. I settled in a forest, on the earth, without boundaries' (Taylor 1982: 178). For Cherry, then, jazz inspired by indigenous folk musics was a form of ecological protest. In its refusal of Western conceptions of tonal harmony, free jazz was a musical emblem of the lack of boundaries that Cherry found in his connection with the natural world. Jazz thus became for him a cure for both physical and spiritual pollution.

The connections that John Coltrane, and later Don Cherry, made with 'world musics' were central to the emerging ecological sensibility among jazz musicians, in that the folk music ideal of non-commerciality has a special affinity with the ethics of radical environmentalism. For multi-instrumentalist Kali Z. Fasteau, as with Don Cherry, ecological awareness can also be expressed in music by fusing styles and instrumentation from the world's traditional musics with the sense of ecstatic, spiritual liberation associated with free jazz. Fasteau believes that, by combining African polyrhythms, Asian microtonal scales and American free jazz, she is exploring new modes of thought. Her album *Oneness* (2003) mixes African djembe with free blowing on soprano saxophone, Arabic mizmar and Nai flute. 'Whale's Reverie' adds bamboo flute and a theremin imitating whale 'song' to Fasteau's mizmar and electronically-treated vocals. Any sense of New Age cliché is dispelled by the exuberance, ferocity and unsettling strangeness of the playing.

In her 'Statement of Purpose', Fasteau appeals to the notion of music as cosmic energy which we explored in Chapter 4. Music, she writes, is 'the most powerful medium of communication, because it transmits the greatest concentration of information through kinetic energy directly to the feeling centres of the heart and spirit as well as the brain, to the very atoms, molecules, cells, and organs of our bodies all moving in their own tandem rhythms. That is why music is such an amazing healing force.' This rhythmical entrainment, she argues, can lead to altered mental states, and ultimately to social change. Musical forms, she writes, 'both reflect and can create and change social structures and behaviour patterns by introducing new thought forms'.

Fasteau argues that modernist techniques of defamiliarisation are vital if music is to foster this higher state of spiritual and political awareness. She continues: 'I blend and juxtapose sounds so as to disturb people's usual ways of thinking and cause them to speculate about life from an entirely different point of view – to realise that separateness is a malevolent illusion, and to warm their spirit' (Fasteau nd). For Fasteau, then, music becomes spiritual nurturance through the very modernist techniques of disturbance and provocation that, for art critic Suzi Gablik and music therapist June Boyce-Tillman, can be bad for ecopsychological health. Artistic challenge and experimentation are for her essential to a developing eco-aesthetics of music. Moreover, as Julian Cowley observes, Fasteau's vision is 'mystical but not otherworld in any evasive sense. It has, she makes clear, an organic and physiological basis, and it bears profound socio-political implications' (2005: 27).

Ecological awareness for Kali Z. Fasteau is necessarily linked to the formal innovations of free jazz and world music. In contrast, tenor saxophonist Sonny Rollins' *Global Warming* (1998) takes a more mainstream approach that fuses bebop with melodic African jazz reminiscent of Abdullah Ibrahim. Rollins saw his album as an extension of his 'Freedom Suite' of 1958, which protested for Civil Rights. Forty years on, he was expressing similar political concerns for the deteriorating environment in titles such as 'Global Warming', 'Mother Nature's Blues' and 'Clear-Cut Boogie'. The Earth is 'a finite planet', Rollins commented on the concept behind the album. 'People have to wake up in time to change this profligate lifestyle which we enjoy' (Europe Jazz Network: nd). Aesthetically, though, *Global Warming* is more therapeutic solace than wake-up call; the catchy, familiar rhythms of South African township jazz replace the challenge of Rollins' earlier work in the 1950s with an affable, if undemanding, groove that emphasises pleasure rather than intellectual challenge.

The German jazz writer Joachim-Ernst Berendt commented in 1983 that in no other profession 'do you find more adherents of the New Consciousness than among musicians'. He cited multi-instrumentalist Collin Walcott, who said that many players of world music anticipate the type of human necessary to survive on a shrinking planet (Berendt 1988: 9). Walcott's group Oregon, which he formed in 1971 with Ralph Towner and Glen Moore, all former members of the Paul Winter Consort, created a quiet, subtle style of chamber jazz that evokes a sense of ecological interconnection without stating it explicitly. Oregon bring together an eclectic range of instruments, including sitar, tabla, classical guitar and English horn, in melodic conversations that can be both contemplative and animated. The band signalled its organicist values by choosing acoustic and non-Western instruments, in album titles such as *Out of the Woods* (1978) and *Roots in the Sky* (1979), and in cover art depicting trees. To use Deleuze and Guattari's terminology, Oregon's drifting, unpredictable melodies and harmonies are more rhizomatic and deterritorialising than arborescent and fixed, and thus further explore the notion that jazz can be a model for benign ecological relationships.

The rest of this chapter explores the work of the two jazz musicians who have been most explicit in developing conceptual links between improvised music and ecological awareness: Paul Winter and David Rothenberg.

## 12.2 'Living Music': Paul Winter

Saxophonist Paul Winter's interest in bringing together improvised jazz with recorded nature sounds began when he heard biologist Roger Payne's recordings of the 'songs' of the humpback whale in 1968. In the liner notes to *Whales Alive* (1987), he recalls his excitement when he attended a lecture given by Payne at Rockefeller University in New York City in 1968: 'I was thrilled by the soulful beauty of these humpback whale voices, in much the same way as I'd been when I first heard jazz saxophonists like Charlie Parker. Listening to the long, complex songs the whales repeat, I was amazed by their musical intelligence and I was shocked to learn that these extraordinary creatures were rapidly being hunted to extinction'. What particularly attracted him to the idea of whales as musicians, he said, was that they are 'such inventive melodists, always taking us in new and unexpected directions.'

All but one of the pieces on *Whales Alive* is derived from melodies 'composed' by the whales themselves. The album was recorded in the Cathedral of St. John the Divine, New York City, the building providing both a resonant acoustic and an appropriate setting for Winter's worshipful celebration of sacred nature. The album was recorded using the flexibility of the church space: keyboardist Paul Halley sat in the organ loft, responding to whale songs played back on a tape recorder operated by Roger Payne, while Winter improvised on his soprano saxophone from the nave, simultaneously listening to the organ and to the whale songs on headphones. Winter explained further: 'I followed Paul's harmonies with my sax and together we extended the whale melodies through improvising, in a way similar perhaps to how the whales themselves gradually change and "grow" their long songs'. *Whales Alive*, he writes in the liner notes, is a 'celebration of optimism, and our song of gratitude to whales for the beauty and wisdom they have brought into our lives'. As in the popular *Free Willy* movies of the 1990s, then, the whale has become a symbol of mutual redemption. As Winter puts it, 'They have helped us mature, and to acquire humility, as our consciousness has grown from "Save the Whales" to "Saved by the Whales".'

Winter's ballad style is melodic, sweetly toned and harmonically familiar. The listener is not pitched into a wild ocean, but rather luxuriates in a nurturing bath of sound. His whales are genteel and spiritual; as he says, it is their 'soulful beauty' that appeals to him, even though, as David Toop notes, one 'eminent marine bioacoustician has suggested that some whale sounds may be non-mystical gastric rumbles' (1995: 241). But Winter's whales are neither bestial nor indecorous, even though the timbre of their sounds is akin to sounds not usually considered beautiful, like a cow's mooing, so that they tend to jar, rather than blend in with, the sweetness of his own music.

Winter's aestheticisation of whale song attempts to lessen the distance between human and animal, so that the animals lose some of their otherness. Indeed, his improvisations with the calls of other wild animals have a similar feeling of nature tamed; what William Cronon calls the 'domesticated sublime' (1996: 75). 'Wolf Eyes', Winter's improvisation with tape recordings of a wolf, released on *Concert for the Earth* in 1987, begins with wolf howls, followed by imitative *glissandi* on his soprano saxophone, lightly underscored by Paul Halley's delicate piano chords. Eugene Friesen's cello amplifies the melancholy. The live recording ends with the audience howling like wolves. Winter notes that it was Stewart Brand, founder of the *Whole Earth Catalogue*, who suggested to him in 1973 that he invite his audience to howl along with the wolf calls. But the attempts at impersonation only add to the melancholy of the piece, connoting not only the sad decline of America's wolf population, but also the unbridgeable gap between human and animal, despite the audience's eager attempt to find its inner wolf.

Winter's vision of a pastoral nature does not appeal to wilderness cultist Lone Wolf Circles, who quipped that the churchiness of Winter's *Gaia* (1982), also recorded in the Cathedral of St. John the Divine, reminded him 'of why I ditched Sunday School, probably the result of too much Merle Haggard and barefoot on my river' (Lone Wolf Circles 1985a: 25). Even though there is a sweet melancholy and piety in Winter's music, however, this has not led him into passivity or complacency. As David Rothenberg notes, Winter was influenced by the photography in the books produced by the Sierra Club, which showed how 'beauty, not only worry and fear, could coax people into the environmental movement' (2008: 40). In 1980, Winter founded Living Music, an organisation which aims to promote music that supports conservationist causes. 'Wolf Eyes' was performed live in front of the General Assembly of the United Nations on World Environment Day in 1985. Winter, then, uses the nurturing quality of his music to add emotional force to his environmental activist concerns.

## 12.3 'Earth Jazz': David Rothenberg

In *The Ecology of Eden*, Evan Eisenberg proposes that the open, improvised quality of jazz makes it a suitable model for benign ecological relationships. He writes: 'How do you collaborate with Gaia if you don't know exactly how she works, or what she wants? You do it, I think, by playing Earth Jazz. You improvise. You are flexible and responsive. You work on a small scale, and are ready to change direction at the drop of a hat. You encourage diversity, giving each player – human or nonhuman – as much room as possible to stretch out.' Eisenberg argues that this dialectic of listening and responding to the natural world finds a happy medium between what he calls the 'Planet Managers', who have an over-aggressive desire to manipulate nature, and the naive 'hands-off' attitude of the 'Planet Fetishers', among whom he includes deep ecologists (1998: 295).

In his book *Sudden Music*, David Rothenberg cites Eisenberg's concept of Earth Jazz, and argues that the unpredictability and non-linearity of jazz improvisation are, for him, the most important features in an ecologically-attuned music. As he puts it, music that is 'poised for the uncertainties of life and able to change its direction in unexpected situations calls attention to the inadequacy of inflexible human plans' (Rothenberg 2002: 97). Rothenberg distinguishes Earth Jazz from John Cage's attitude of acceptance towards all of the sounds in his environment. 'It may be', he muses, 'that the most earth jazz can teach us it to notice things, to listen to interactions: not simply to smile, enjoy, and take it all in as Cage and the soundscapers would have us do, but really to find a way to fit in – musically, culturally, individually, and collectively, to respond, to wend a way inward and outward at once' (202).

Rothenberg has built Cage's naturalist aesthetics and the philosophy of Arne Naess into an explicitly deep ecological theory of music. He has high ambitions for his music, wanting nothing less than 'to save this Earth through music' (175). In order to achieve this, musicians must learn to 'work like nature'; that is, to 'figure out just how nature operates aesthetically' (190). According to this Cagean principle, 'something can be good if it *sounds like* nature. And if it's especially good, it will change the way we hear nature, define nature, and then live in nature. Hopefully the trajectory of Western culture has taught us to hear more, not less, and to hear enough to make us question the whole course of that culture. By listening, by dancing, by grooving, music itself can become an agent for change' (97).

At the centre of Rothenberg's vision for music is Naess's phenomenological theory of 'concrete contents'. Rejecting a firm distinction between subject and object, Naess draws on the seventeenth-century philosophical distinction between primary, secondary and tertiary qualities. The primary qualities of an object include its size, shape, number, motion and rest. Its secondary qualities include colour, taste and heat. Tertiary qualities are feelings, sentiments and emotions. Crucially for his philosophy of deep ecology, Naess denies that tertiary qualities are subjective. Instead, he considers them to be as inherent in the object itself as the primary and secondary ones. Naess thus sees no distinction between the world as it actually is and the way we experience it.

In this version of Gestalt psychology, Naess argues that the human subject that perceives the world, and the objects that make up that world, both exist in an interactive relationship. From this, he deduces two important points. Firstly, that to consider the world as made up of discrete 'things' is to abstract from the interactive relationality on which reality is based. Secondly, this relationality means that human subjects do not merely project emotions or ideas onto a world that is external to them. As he puts it, '*Gestalts bind the I and the not-I together in a whole*. Joy becomes, not *my* joy, but *something joyful* of which the I and something else are interdependent, non-isolatable fragments. "The birch laughed/with the light easy laughter of all birches …" This gestalt is a creation which may only incompletely be divided to give an I which projects laughter into a non-laughing birch tree. The glorification of conventional "scientific" thought leads to the ridicule of such creations. It tears gestalts asunder' (Naess 1989: 60-1) For Naess, then, abstract forms of knowledge, epitomised by modern science, should not be deemed superior to direct, spontaneous human perception.

In his essay '"No World But in Things"': The Poetry of Naess's Concrete Contents', Rothenberg gives his own version of this deep ecological position. For Naess, he writes, 'the emotions we sense there in nature are not mere projections, or whims of different perspectives, but an actuality that can't be denied'. This epistemological shift, he continues, 'might be the birth of a less arrogant way of placing human beings into an equally experiencing surrounding, also a subject, not only an object' (Rothenberg 2000: 153).

But there are problems with Naess's philosophy, and therefore with Rothenberg's application of it to music. As Eccy De Jonge points out, some subjective perceptions are simply wrong, and therefore should be attributed to the misapprehensions of the perceiving subject, rather than to the object itself. Hence, the distinction between subject and object is upheld. Moreover, Naess's philosophy is an extreme form of anthropocentrism. 'Strange that the deep ecologist, who wants to maintain a theory of intrinsic value', writes De Jonge, 'should dismiss the thing-in-itself so readily. While Naess's ontological analysis helps to eliminate the isolated ego, the relational field *at the same time* eliminates the possibility that things contain an *intrinsic value*' (italics in the original) (De Jonge 2004: 43).

In applying such deep ecological principles to music, Rothenberg's philosophical method is speculative rather than analytical: he tends not to address objections to his *a priori* assumptions in order to produce a better theory. Instead, he develops Naess's deep ecology into a rhapsodic celebration of the potential of music to heal ecological problems. He explains the naturalist basis of this aesthetic position as follows: 'Although it is a challenge, the goal can be to make something that helps humanity *fit in to nature*. This may be easy for other life forms, but it is hard for us, caught as we are in our own arbitrary rules that simplify the world beyond belief: rectangles; four-four time; beginnings, middles, and ends. The world out there admits none of this. It is a booming, buzzing confusion, and will stay that way long after we give up trying to reduce it to principles and laws' (Rothenberg 2002: 129).

The quotation from William James in the previous sentence identifies Rothenberg as part of an American philosophical tradition of radical empiricism, whose belief in what James called 'a world of pure experience' anticipated the desire for unmediated experience in both phenomenology and deep ecology (Kuklick 2001: 171). Rothenberg applies this distrust of abstraction to his aesthetics of music. Four-four time is an arbitrary rule, he asserts, that merely simplifies our experience of the world.

In practice, Rothenberg's own music does tend to avoid a regular four-four beat. The synthesized drum track on 'In the Rainforest' changes metre, while the clarinet drifts meditatively across the bar lines. Moreover, although he uses beats generated by computers, he avoids the repetitive, looped 'four-to-the-floor' beat familiar in many forms of electronic dance music (Rothenberg 2002). Nevertheless there is an inconsistency in Rothenberg's attitude to duple time. A few pages earlier in *Sudden Music*, he writes approvingly about music 'that seems to *live* just by the way it moves, and by how it draws the listeners and players closer together around a common, organic pulse [...] It's the thrum of life, the catchy beat, the pattern that the drummer and woodpecker can share' (Rothenberg 2000: 96). Yet the catchy, common 'pulse' of most dance music is based on the very four-four metre he later dismisses as an unnatural simplification of nature. It is, rather, Rothenberg's deep ecological distrust of abstraction that leads him to dismiss the four-four beat as 'arbitrary'.

This distrust of abstraction also informs his attitude to environmentalism. There is, he writes, 'a level of understanding that takes a step back from the immediacy of relating with the world: deciding how much land is worth in monetary terms, or discovering the chemical formula that constitutes soil or the cell pattern of wood. These are *abstract structures*, once removed from the concrete relational contents of the nature that includes our ability to know it'. He continues: 'The implications of this view for tactical, pragmatic environmentalism are clear: if a defender of a mountain sees it as beneficent, glorious, sublime, then that value is as *primary* as the value of the mineral deposit that might lie beneath the summit, ready to be excavated and converted into cash, even though it is harder to quantify' (153).

Rothenberg here bases his defence of radical environmentalism on Naess's philosophically flawed theory of concrete contents. But the argument that aesthetic and emotional responses to the natural world are as important as scientific views, in this case from economics and chemistry, does not require Naess's idealist theory for its justification. People can consider their emotional responses to the natural world important, even if they are only projections onto an exterior object. Links between music and environmentalism can be justified socially and politically, without requiring arguments from nature or from questionable ecophilosophy. Moreover, the intellectual processes of abstraction opposed by deep ecologists like Naess and Rothenberg are vital to the very science on which they base their knowledge of the world's ecological problems. As science writer Stephen Budiansky explains, with a musical analogy that ironically turns Rothenberg's argument on its head, 'It makes as much sense to denounce a scientist's use of mathematics to build up a picture of nature, as it would to criticise Beethoven for having composed a symphony out of individual musical notes' (1996: 25).

## 12.4 Jamming with nature

Rothenberg is one of a number of musicians whose desire to ground their music in the natural world has led them to play music directly to, or, as they would say, 'with' animals. Jim Nollman has made the most systematic attempt to communicate with animals through music, forming the organisation Interspecies Communication in 1984 to fulfil this ambition. In his books, he describes playing music to a pod of orcas, and eventually managing to elicit a vocal response from them. He sees this as a tentative first step towards establishing a new relationship of communion with nature (Nollman 1999: 207-8; 2002: 129-37).

For Rothenberg, jamming with nature in live performance is preferable to using taped samples of natural sounds in studio recordings, as in the ambient electronica discussed in the previous chapter (Rothenberg 2002: 95). He believes that music has the capacity to reconnect us with animals, and even plants, because it is a non-verbal art form, and therefore bypasses the alienation from nature that arises from the human capacity for language. Music, he writes, 'is a form of communication that reaches out to the rest of the world. Animals and plants may also grasp it, unlike all that talk, which is utterly human and nothing more' (164).

Playing live music with birds, Rothenberg writes, is '*not* about expressing the self, but about expressing something larger than the self, a way toward fitting in with the natural world that belies the human sense of separateness, inadvertence, and doubt' (186). His book *Why Birds Sing* explores the hypothesis that birds sing not only for reasons of territory or reproduction, but for 'fun' and 'joy' (Rothenberg 2005: 97). Rothenberg also presents his recordings as evidence of mutual interaction between bird and human. 'White Crested Laughter' features Rothenberg himself on clarinet and Michael Pestel on flute, improvising with a white-crested laughing thrush at the National Aviary in Pittsburgh in March 2001. The piece begins with the players directly imitating the bird's calls, before they move into more elaborate improvisation. At first, the thrush is silent, then it starts to sing. The singing at first appears to be independent of the instrumentalists, but towards the end of the eight minute track, it is possible that the bird is responding to what it hears in a more direct way (Rothenberg 2005).

Yet Rothenberg's belief that when he plays music to birds, the latter participate with him on an equal basis, is surely questionable, as the precise nature of whatever 'communication' is established remains unknown, and whether genuine reciprocity has actually taken place is at least doubtful. The deconstructionist philosopher Paul de Man wrote of John Keats' 'Ode to Melancholy', that Man 'feels envy for the natural creature that is the direct emanation of Being' (1983: 238). Playing music to animals is for Rothenberg, as for Jim Nollman, an attempt to close this ontological gap, to overcome the separation of human beings from the rest of nature that more dualistic philosophies see as inevitable. Both cite the religious beliefs of Native American cultures to justify the ecological beneficence of such desires for unity with nature. Nollman quotes Luther Standing Bear on the unity of human beings with nature (2002: 163). Rothenberg quotes Nalungiaq, a Nestilik Eskimo, talking of an original time when humans and animals spoke the same language. 'Not only have we lost that language,' he adds, 'we can barely imagine what it might be. Words are not the way to talk to animals. They'd rather sing with us – if we learn their tunes without making them conform to ours' (Rothenberg 2002: 87).

The authority imputed to Native American religions apparently guarantees the reciprocity of these human-animal interactions. However, although such performances are intended to respect animal subjectivity, they probably remain an exercise in Romantic wishful thinking. Rothenberg writes of playing music with animals: 'I'll happily romanticise what other creatures might want from me, and seek out the tools that let us loose from our humanity, allow us to speak with accustomed wildness, get us into the wind and the seas, as we find our way to hollows in the soundscape that await our ideas' (186). Ultimately, he can blithely dismiss any epistemological problems with his philosophical claims by applying Arne Naess's theory of concrete contents, which, as we saw earlier, rejects the realist idea that human beings are subjects who project emotions and ideas onto natural objects that are external to them.

To suggest philosophical problems with Rothenberg's theorising of music and nature is not to argue that music is necessarily exclusive to human cultures. The developing field of zoomusicology explores the possibility that the best model for animal communication is music rather than speech (Wallin et.al 2000). Accordingly, Deleuze and Guattari apply their theories of music equally to birds and to the French composer Olivier Messiaen (1987: 316-7). To improvise, they write, is 'to join with the World, or meld with it. One ventures from home on the thread of a tune' (344). However, as Alan Durant argues, any ideas about 'social values and models of interaction' achieved either through playing or listening to improvised music are 'made in advance or held by presumption'. Improvised music, he concludes, 'merely dramatises and endorses an already established framework of values, rather than constructing or investigating them. To follow this line of thinking is to move back from claims that improvisation has a unique investigative or therapeutic potential to the more orthodox view that music-making *represents* particular ideas, feelings or values' (Durant 1989: 275). According to this argument, then, music is ultimately an aesthetic rather than a philosophical experience, so that listeners who find a link between improvised music and ecological thinking are drawing on their prior knowledge of the latter, rather than finding a quality inherent in the music itself. But this is not something to mourn; rather, it is sign of the inventiveness of the human imagination and the importance of music as emotional and cognitive play.

# Afterword

This book has explored a wide range of provisional answers given by musicians, either explicitly or implicitly, to two fundamental questions: 'What sort of music?', and 'What sort of ecology?' We have seen that conceptual links between music and ecological thought include the placement of explicitly environmentalist subject matter in the lyrics of songs, as well as notions of 'eco-listening' that theorise the way music works on its listeners. Faced with diverse claims, it is difficult to argue that one form of music is more 'natural' or 'ecological' than any other. Many different styles of music have their advocates as means of fostering environmental awareness. Indie rock songs like R.E.M.'s 'Cuyahoga' use modernist strategies of defamiliarisation to create openness and ambiguity; in contrast, the proletarian realist tradition of American folk music, led by Pete Seeger, aims for clarity and unanimity of audience response.

Aesthetically, the popular music discussed in *The Jukebox in the Garden* raises two key issues: the role of sentiment and of didacticism in music. For Timothy Morton, a sentimental approach to nature is complicit with the consumerist appropriation of nature under capitalism. Yet the boundaries between sentimental kitsch and 'high environmental art' require constant policing; indeed, the latter may itself be 'just a disavowal of kitsch that is, uncannily, its inner essence' (Morton 2007: 154). Moreover, kitsch, because it is 'unashamed about its status as a mass-produced commodity', may possess a critical function that more refined high art lacks (151).

In any case, unreflective sentimentality in environmentalist art may have had a more radicalising effect than Morton acknowledges. As David Whitley argues in relation to Disney's nature movies, there is a convincing argument that the 'Bambi syndrome' or 'Disneyfication' of nature has been influential in rallying support for environmental causes, including radical and deep ecological ones (2008: 3-4). The sentimentality of popular music, typified by John Denver and other exponents of the pastoral mode explored in this book, may play an equally important role in environmentalist advocacy. Historians of nineteenth-century reform movements such as abolition and the protests against American Indian removal have written of the value of sentimentality in the development of a politics of sympathy for the oppressed, where it provided a strategy for white women to identify, however problematically, with their racial and class 'Others' (Chapman and Hendler 1999: 6). From this perspective, when popular singers risk sentimentality by singing the praises, or mourning the loss, of a particular place, landscape or wild animal, they are encouraging their audience to connect emotionally with the natural world in a way that may lead to environmental awareness and action. As Ernst Bloch noted, even the naive dreams of an 'abstract utopia' are preferable to resignation and indifference; Leo Marx's 'simple pastoral', however sentimental and kitsch, can, as Laurence Buell argues, articulate dissatisfaction with the status quo of modern, industrial capitalist society.

Problems arise, however, when a sentimental view of nature is confused with a scientific one. Indeed, some scientific ecologists have argued that sentimentality in relation to nature can harm, rather than encourage, effective policies of environmental conservation and restoration, when it contradicts the scientific evidence on which those policies need to be based (Budiansky 1996: 3-26). Despite these reservations, however, sentimental attachments to the natural world are, from both an experiential and a political point-of-view, as important as scientific analyses, and are a vital part of environmental concern.

Like sentimentality, didacticism is another crucial aesthetic issue for any study of popular music and environmentalism. Jerry Garcia of The Grateful Dead was aware of the aesthetic problem of trying to represent global environmental issues in music. He told an interviewer: 'We are trying to make things groovier for everybody so more people can feel better more often, to advance the trip, to get higher, however you want to say it – but we're musicians, and there's just no way to put the idea "save the world" into music' (Troy 1994: 114). But the relatively modest ambitions for rock music expressed by Garcia here ran counter to the wider notion that the lyrics of rock records can provide a serious critique of contemporary society, including its environmental ills. Yet Simon Frith argues that rock musicians in the sixties were misled by a sense of their own importance as artists, to the point where songs 'which should have worked as a fleeting subversion of moods became too didactic to be used by anyone'. Ironically, then, he argue that songs 'that took it for granted that they didn't matter registered much more precisely than progressive rock the essential mood of the 60s, the sense of change and possibility' (1984: 68).

While Frith has a point, his conclusion, that 'the only political issues on which popular music has any bite' are 'issues of pleasure, escape, banality', unnecessarily limits the range of popular music's concerns, and plays down the art of the songwriters and performers discussed in this book, who have taken on social issues, including threats to the environment, with varying degrees of subtlety, intelligence and emotional depth. A more nuanced perspective is given by rock critic David Downing, who also argues that didacticism is aesthetically problematic in rock music, which is 'not a medium for making statements' (1976: 29). It is for this reason, he notes, that 'the growing ecological crisis has received little more shrift than the Bomb'. Rock music works best, Downing writes, 'when music and lyrics combine for three minutes or three seconds, to produce a flash of consciousness, not by any means a necessary articulated consciousness, of the relationship between the one and the world [...] Rock does not explain; it compresses and connects.'

By 'the one', Downing means the self: a good rock song will connect abstract issues to a sense of concrete, individual experience. Other aesthetic criteria for a successful popular song can include a seeming unity of words and music, brevity and concision of expression, and use of obliqueness and analogy rather than of overt didacticism. Randy Newman's 'Burn on Big River', on his 1972 album *Sail Away*, may stand as an example of analogical connection that avoids the bathos into which an overly didactic approach to environmental concerns may descend. The song concerns the chemical fire on the Cuyahoga River in Cleveland, Ohio in June 1969, which was also the subject of R.E.M.'s 'Cuyahoga'. In an apocalyptic landscape, a red moon rises over an oil barge on the river. Cleveland is a 'city of light', but the price paid for this technological magic is environmental degradation. Newman's piano accompaniment suggests a gospel hymn, as the chorus indicts human transgressions of nature: 'the Lord can make you overflow / But the Lord can't make you burn'. The imagistic compression and obliqueness of Newman's lyrics combine with his understated vocal delivery to make 'Burn on Big River' topical without trivialising the issues it deals with.

Another useful critical case study for an eco-aesthetics of popular music is Michael Jackson's 'Earth Song' (1995). Ever since sixties rock bands like The Grateful Dead and Jefferson Airplane signed to the big record labels of their day, rather than to local independent record companies, countercultural protest, including against threats to the environment, has been sold as a commodity by large capitalist corporations (Harron 1988). Jackson's 'Earth Song' probably encapsulates more than any other the way that the newly globalised music corporations, in this case Sony, are providing a forum for environmentalist protest, if it is suitably vague and unfocussed. Yet if Jackson's lyrics reduce environmental apocalypse to a series of predictable questions ('What about the rain?', 'What have we done to the world?'), his vocal performance adds musical interest, as he embellishes this mounting series of rhetorical questions with whoops, sighs and groans, while the gospel call-and-response arrangement drives the song towards an impassioned, if melodramatic climax. In terms of environmental ethics, the overtones of Messianism in Jackson's performance in the video, and at the 1996 Brit Awards in London, were more troubling: American pop star as lone saviour of the Earth's endangered peoples and animals.

When ethics are conceptually separated from aesthetics, on the lines suggested by aesthetic philosopher Berys Gaut in the Introduction to this book, critical appreciation of a piece of music, such as 'Earth Song', does not depend on agreeing with the ideas it expresses. I have argued against many of the assumptions of deep ecology in this book, but do not believe that his sympathy for such ideas, however idiosyncratic, makes the music of Captain Beefheart any less enjoyable or culturally important. The aesthetic pluralism at the centre of the book's approach to music has thus enabled artistic value to be found in a wide range of music. Often this value can be found in small details: Pete Seeger's delicate banjo picking underneath the lilting melody of 'My Dirty Stream (Hudson River Song)', Gil Scott-Heron's vocal riffing on 'South Carolina (Barnwell)', or the exuberance of Kali Fasteau's eclectic jazz arrangements.

Further research is needed in reception studies to investigate how particular pieces of music have actually affected listeners, and whether they have played a part in organisations or subcultures involved in environmental activism. Such studies may establish how the music I have discussed in this book has been listened to by listeners other than myself. Although such studies are beyond the scope of this book, it is important to note that any critical move from ideological analyses of works of art, such as those attempted here, to assumptions about their effects on real audiences should not be too rapid. As Herbert Gans argues, claims that popular culture, by offering 'contrived solutions to insoluble problems', encourages its audience to escape their problems in real life, perhaps make too simple a link between an art work and its effects on thought and behaviour (1974: 35). Therapeutic, 'feelgood' music, like Sonny Rollins' *Global Warming* or Mickey Hart's *Planet Drum,* is not necessarily any more or less effective in encouraging social change than the modernist provocations of Kali Fasteau or Frank Zappa.

The popular music featured in this book raises questions regarding American ethical, philosophical and political responses to an age of environmental crisis. Throughout history, human beings have attributed many different functions to music, from giving pleasure, to providing moral and political education, to possessing magical and supernatural powers of transformation (Fubini 1987: 8). The music discussed here has variously embraced all of these functions. Yet the question of how music relates to ecophilosophy ultimately depends on the type of philosophy one chooses to adopt. Music is not analytical philosophy, because it is largely incapable of producing propositional knowledge about the world. Using as his example Beethoven's imitation of bird calls in his Sixth Symphony, 'The Pastoral', musicologist Peter Kivy concludes that the propositional content of the symphony is limited to, 'Here is a cuckoo; it sounds a descending major third' (1984:157). This failure to produce propositional knowledge leads analytical philosopher Philip Alperson to argue that music has therefore only played two peripheral roles in philosophy. Firstly, it has had a 'genetic' role in inspiring philosophical concepts or theories, as in the famous stories of Pythagoras formulating his philosophy on hearing the sounds of a blacksmith, or Einstein attributing his breakthrough in modern physics to his violin lessons. Secondly, it has had an 'illustrative' role, as in the philosopher Henri Bergson's use of melody to illustrate his theory of time, or the musical analogy that contemporary 'string theory' draws upon to explain the fundamental constitution of the sub-atomic world (Alperson 1987: 203-4). After relegating music to this marginal role in philosophy, Alperson asserts that 'the heyday of musically-based speculation has passed'. Music, he writes, 'may never again enjoy the centrality in that regard that it did in the ancient and medieval worlds' (203). Yet many of the musicians and ecophilosophers featured in this book would beg to differ with this claim.

The philosophical basis for many of the ideas explored in this book is an understanding of music, and of art in general, as an important knowledge-producing activity in its own right. Philosopher Eileen John starts from the same premise as Alperson: that art does not meet the requirements for the production of knowledge, if the latter is narrowly defined as true, justified belief. However, she argues that 'We need a theory of knowledge which embraces such things as knowing how to perceive, imagine, and feel aptly, and knowing what a certain experience is like'. She goes on to give an evolutionary explanation for art's potential to integrate 'pleasure-seeking into cognitive activity' (John 2001: 332). As Wittgensteinian philosopher Oswald Hanfling argues, 'learning is not always a matter of acquiring new information. A work of art, like a work of philosophy, can provide *insight* into the human condition without informing us of facts of which we had been ignorant' (2001: 90).

As an art form, music can nurture our imaginative, emotional and spiritual responses to the natural world. In doing so, it calls upon both analytical and intuitive modes of thought, and thereby extends human rationality, rather than conflicts with it. The humanist Marxists Ernst Bloch, Theodor Adorno and Henri Lefèbvre all argued for this expanded notion of rationality. Adorno wrote that 'every art which deserves attention approaches the aim of rationality by its very structure, and tends more and more toward "knowledge"' (Buck-Morss 1977: 123). Bloch argued that the anticipatory element in utopian art, including music, acts '*as a directing act of a cognitive kind*' (1995: 12). Henri Lefèbvre wrote that there is a 'vast store of non-formal knowledge embedded in poetry, music, dance and theatre. This store of non-formal knowledge (*non-savoir*) constitutes a potential true knowledge (*connaissance*)' (1991: 407).

In contrast to these materialist thinkers, deep ecologists and New Agers tend to be more ambivalent about the value of rationality. Although they often pay lip service to the notion that analytic and intuitive modes of thought should be integrated, in practice they tend to denigrate the former, as we saw in Fritjof Capra's Preface to Berendt's book *Nada Brahma* quoted in Chapter 12. This Romantic anti-rationalism has been seen throughout this book. New Age neo-pagan

Lone Wolf Circles claims that music can restore the listener to oneness with the natural world only by silencing intellect and verbal language: 'Soothing and seductive, or loud and powerful', he writes, 'it is the function of music to wrest your attention from trivial intellectualisation. Like a clap of thunder or a sexy purring in the ear, music must leave the omnipresent commentator of the mind speechless' (Lone Wolf Circles 1991: 53). The sixties rock counterculture also tended towards this critique of rationalism. In The Grateful Dead's 'Dark Star', 'reason tatters' to reveal a 'transitive nightfall of diamonds' full of wonder and potentiality.

The limitations of this view of music as a non-verbal and non-rational experience have been exposed by musicologist Victor Zuckerkandl. Music, he wrote, does not mark 'a reversion to earlier, pre-linguistic, prerational stages of development', but is instead about 'going forward to a broadened rationalism'. He thus rejected the idea that music (or what he called 'the tone') 'feeds the irrational element as the word feeds the rational element. Such either/ors are false and superficial'. Rather, music 'does not thrive at the expense of rationality. Music originates, grows, and reaches its culmination within human rationality, together with it, not outside or against it' (Zuckerkandl 1973: 76).

The approach adopted in this book has been sceptical of the idealist assumptions of both deep ecology and New Age ecophilosophy, and has defended a version of political ecology based on critical realist notions of science and self-reflexive rationality. Scientific accounts of music may seem less glamorous than the more mystical accounts favoured by New Age thinking, but they provide a more empirical understanding of the effects of music on the human mind-body. Scientific research into music and listening is ongoing and provisional, but, like science as a whole, does not reveal a world any less full of wonder than that conjured up by idealist eco-mysticism.

Nevertheless, a defence of rationalism and materialism does not have to conflate spirituality, understood as transcendental mysticism, with broader notions of individual and collective well-being which are essential to a wider sense of ecological health. Political scientist Andrew Dobson writes that the metaphysics of oneness is 'a mystical answer to a problem whose rational solution needs to be found in material human practice rather than in contemplation' (1993: 193). Yet the role of 'contemplation' as a method of extending human rationality, rather than rejecting it, should also be considered. However, when New Agers and deep ecologists insist that nature mysticism and a return to the 'sacred' in nature are fundamental to social and environmental renewal, or to living an ecologically sounder life, they risk becoming a sanctified elite of true believers who exclude those who cannot make the necessary leap of faith. If the call for environmental reform (or revolution) is to be heeded more widely, it will require alliances between people with many different ideas and beliefs, from Buddhists, Muslims and Christians to agnostics and atheists, and cannot merely be the preserve of a self-elected countercultural or religious minority.

What can unite people across such cultural divides is, of course, music. This utopian potential in music has been the thread that unites many of the musicians in this book, from Pete Seeger and David Rovics to Paul Winter and Brian Transeau. Such eco-utopianism has its limitations. As anthropologist Martyn W. Lewis argues, the challenge for environmentalism 'must be more than to criticise society and imagine a blissful alternative. On the contrary, the movement must devise realistic plans and concrete strategies for avoiding ecological collapse and for reconstructing an ecologically sustainable economic order' (Lewis 1992: 250). Yet the crucial question of agency, of how an artistic subculture of musicians and their audiences might connect with wider social and political movements, including those in defence of the environment, remains vague in many of the utopian writings on music that feature in this book. In any case, as Ruth Levitas observes, it is 'something of a last resort to require artistic production to be *the* source of transformative energy in society' (1990: 196).

Art, though, can provide a space for imaginative and speculative play. Music is obviously not a solution to environmental problems in and of itself. Yet the idea that it prefigures a better society, including a better relationship between human beings and the natural world, is an attempt to account for the profound effect that music has on its listeners. Ernst Bloch's notion of the 'surplus' of 'hope-material' in music remains an endorsement of music's enduring utopian value, beyond its status as a commodity. Charles Keil's solution to the problem of commodification in music is to encourage mass participation in live music-making. Yet the commodity status of Mickey Hart's *Planet Drum* does not undermine his claims for its ecological role: its use value is in excess of its exchange value. Music will not bring down the Walls of Jericho on its own, still less curdle milk or restore ecological health, but it can play a vital role in forging group solidarity and identity, and in providing hope, pleasure and spiritual uplift for its listeners. To give the last word to Ernst Bloch, the relation of music to the future is 'seismographic', in that it 'reflects cracks under the social surface, expresses wishes for change, bids us to hope' (1995: 1088).

# Recordings cited


10,000 Maniacs. 1989. *Blind Man's Zoo*. Elektra.
Arrested Development. 1992. *3 Years, 5 Months and 2 Days in the Life Of …* Cooltempo.
Atom Heart, Tetsu Inoue and Bill Laswell. 1995. *Second Nature*. Submeta.
Axelrod, David. *Earth Rot*. Stateside. 2001.
Band, The. 2000. *Music From Big Pink*. Capitol.
Banhart, Devendra. 2004. *Nino Roja*. XL.
Beach Boys, The. 1971. *Surf's Up*. Capitol.
Beaver and Krause. 1970. *In A Wild Sanctuary*. Warner Bros.
Berry, Chuck. 1993. *The Blues Collection*. Orbis.
Browne, Jackson. 1973. *For Everyman*. Asylum.
– *Late For The Sky*. 1974. Elektra/WEA.
BT. *ESCM*. 1997. Perfecto / Kinetic / Reprise.
Byrds, The. 1965. *Mr. Tambourine Man*. CBS.
– 1966. *Fifth Dimension*. CBS.
– 1967. *Younger Than Yesterday*. CBS.
– 1968. *The Notorious Byrd Brothers*. CBS.
Canned Heat. 1990. *The Best of Canned Heat*. Capitol.
Captain Beefheart and the Magic Band. 1967. *Safe As Milk*. BNG/Buddah.
– 1968. *Strictly Personal*. Liberty.
– 1970a. *Trout Mask Replica*. Reprise.
– 1970b. *Like My Decals Off, Baby*. Rhino.
– 1971. *The Spotlight Kid*. Reprise.
– 1972. *Clear Spot*. Reprise.
– 1978. *Shiny Beast* (*Bat Chain Puller*). Virgin.
Carlos, Wendy. 1998 *Sonic Seasonings*. Serendip.
Cash, Johnny. 1962. *Blood, Sweat and Tears*. Columbia/Legacy.
Chapman, Tracy. 1995. *New Beginning*. Elektra.
Charles, Roosevelt, assisted by Otis Webster. 1978. *Mean Trouble Blues*. Vanguard.
Charlie Daniels Band, The. 1985. *Me and the Boys*. Epic.
Cherry, Don. 1971. *Orient*. Affinity.
Coltrane, John. 1965. *Kulu Sé Mama*. Impulse!
Creedence Clearwater Revival. 1990. *Chronicle, Vol. 1*. Fantasy.
Crosby, Stills and Nash. 1969. *Crosby, Stills and Nash*. Atlantic/WEA.
Crosby, Stills, Nash and Young. 1970. *Déjà Vu*. Atlantic/WEA.
– 1971. *Four-Way Street*. Atlantic/WEA.
Crosby, David and Graham Nash. 1975. *Wind on the Water*. Reactive.
David Thomas and the Pedestrians. 1981. *The Sound of the Sand and Other Songs of the Pedestrian*. Rough Trade.
David Thomas and Two Pale Boys. 2001. *Surf's Up!* Glitterhouse.
Dead Kennedys. 2001. *Fresh Fruit For Rotting Vegetables*. Cherry Red.

Denny, Martin. 1996. *The Coconut Cream of Martin Denny.* Elite.
Denver, John. 1971. *Poems, Prayers and Promises.* RCA.
– 1972a. *Rocky Mountain High.* RCA.
– 1972b. *Farewell Andromeda.* RCA.
– 1974. *Back Home Again.* RCA.
– 1975. *Windsong.* RCA.
– 1980. *Autograph.* RCA.
– 1982. *Seasons of the Heart.* RCA.
– 1983. *It's About Time.* RCA.
– 1986. *One World.* RCA.
Di Micele, Alice. 1996. *Too Controversial.* Alice di Micele.
Dr. Octagon. *The Return of Dr. Octagon.* Ocd.
Duke Ellington. 2006. *New Orleans Suite.* Atlantic.
Dylan, Bob. 1963. *The Freewheelin' Bob Dylan.* Columbia.
– 1964. *The Times They Are A' Changin'.* Columbia.
– 1965. *Highway 61 Revisited.* Columbia..
– 1966. *Blonde on Blonde.* Columbia.
– 1968. *John Wesley Harding.* Columbia.
– 1969. *Nashville Skyline.* Columbia.
– 1970a. *Self-Portrait.* Columbia.
1970b. *New Morning.* Columbia.
– 1975. *The Basement Tapes.* Columbia.
– 1976. *Desire.* Columbia.
– 1983. *Infidels.* Columbia.
– 1985. *Biograph.* Sony.
Eagles, The. 1976. *Hotel California.* Elektra/WEA.
Fasteau, Kali Z. 2003. *Oneness.* Flying Note.
Flying Burrito Brothers. 1969. *The Gilded Palace of Sin.* Edsel.
Franti, Michael and Spearhead. 2003. *Everyone Deserves Music.* Parlaphone.
Fugs, The. 2006. *Electromagnetic Steamboat.* Rhino.
Gaye, Marvin. 1971. *What's Going On.* Motown.
Grandaddy. 2003. *Sumday.* VZ Records.
Grandmaster Flash and the Furious Five. 1982. *The Message.* Sugar Hill Records.
Grateful Dead, The. 1967. *The Grateful Dead.* Warner Bros.
– 1968. *Anthem of the Sun.* Rhino/WEA.
– 1969. *Live/Dead.* Warner Bros.
– 1970. *Workingman's Dead.* Warner Bros.
– 1987a. *In the Dark.* Rhino.
– 1987b. *So Far.* Virgin Music Video. (Video).
Griffith, Nanci. 1987. *Lone Star State of Mind.* MCA.
Guthrie, Woody. 1982. *The Very Best of Woody Guthrie.* MCCD.
Halpern, Steven. 2001. *Chakra Suite: Music for Meditation, Healing and Inner Peace.* Inner Peace.
Harris, Emmylou. 1992. *At the Ryman.* Reprise/WEA.
Hart, Mickey. 1991. *Planet Drum.* Rykodisc.
Hassell, Jon. 1983. *Magic Realism: Aka-Darbari-Java.* EEG.
Henley, Don. 1984. *Building the Perfect Beast.* Warner Bros.
– 2000. *Inside Job.* Warner Bros.

Hinojosa, Tish. 2003. *The Best of Tish Hinojosa Live*. Rounder.
Holiday, Billie. 2001. *Lady Day: The Best of Billie Holiday*. Sony.
Hooker, John Lee. 1964. *The Vee Jay Years 1955-1964*. Charly.
Hudak, John. 1997. 'Sketch of a Field' on Various Artists. *Scatter*. Ash International.
– 1998. *Brooklyn Bridge*. Shirocoal.
– 2001. *Highway*. Edition.
Jackson, Michael. 1995. *HIStory: Past, Present and Future, Book 1*. Epic.
Jefferson Airplane. 1967a. *Surrealistic Pillow*. RCA.
– 1967b. *After Bathing At Baxter's*. RCA.
– 1968. *Crown of Creation*. RCA.
– 1969. *Volunteers*. RCA.
Kantner, Paul. 1970. *Blows Against the Empire*. RCA.
Kantner, Paul and Grace Slick. 1971. *Sunfighter*. RCA.
Lansky, Paul. 1993. *Homebrew*. Bridge.
McDonald, Country Joe. 1976. *Paradise with an Ocean View*. Fantasy.
McFarland, Gary. 1968. *America the Beautiful, an Account of its Disappearance*. DCC Jazz.
McGuire, Barry. 2000. *Eve of Destruction*. MCA.
Mamas and the Papas, The. 1990. *The Papas and the Mamas*. MCA.
Memphis Minnie. 2003. *Me and My Chauffeur*. Proper Records.
Merchant, Natalie. 2001. *Motherland*. Elektra/WEA.
Miller, Frankie. 1961. 'The Blackland Farmer' on Various Artists. *The Country Music Hall of Fame*. Columbia.
Mitchell, Joni. 1970. *Ladies of the Canyon*. Reprise/WEA.
Mos Def. 1999. *Black on Both Sides*. Rawkus.
New Riders of the Purple Sage. 1971. *New Riders of the Purple Sage*. Columbia.
Newman, Randy. 1970. *12 Songs*. Reprise/WEA.
– 1972. *Sail Away*. Warner Bros/WEA.
Newsom, Joanna. 2005. *The Milk-Eyed Mender*. Drag City.
Nitty Gritty Dirt Band, The. 1986. *Twenty Years of Dirt*. Warners.
– 2002. *Will The Circle Be Unbroken*. Capitol.
Nollman, Jim. 1989. *Orca's Greatest Hits*. Interspecies.
Ochs, Phil. 1964. *All the News That's Fit to Sing*. Hannibal.
Oregon. 1978. *Out of the Woods*. Discovery.
– 1979. *Roots in the Sky*. Discovery.
Owens, Buck. 1992. *The Buck Owens Collection*. Rhino/WEA.
Parsons, Gram. 1990. *GP/Grievous Angel*. Reprise.
Paul Winter Consort. 1987. *Concert For The Earth*. Living Music.
Payne, Roger. 1971. *Songs of the Humpback Whale*. Living Music.
Paxton, Tom. 1999. *The Best of Tom Paxton*. Elektra.
Pere Ubu. 2005. *The Modern Dance*. Essential.
Pixies, The. 1989. *Doolittle*. 4AD.
Prine, John. 1971. *John Prine*. Atlantic/WEA.
Pride, Charley 1970. *From Me To You*. RCA/Victor.
Psychick Warriors ov Gaia, The. 1995. *History of Psychick Phenomena*. KK Records.
Quicksilver Messenger Service. 1971. *What About Me*. EMI.
Reed, Lou. 1989. *New York*. Rhino.
R.E.M. 1993. *Life's Rich Pageant*. IRS Records.

Reynolds, Malvina. 2000. *Ear to the Ground.* Smithsonian Folkways.
Richard Hell and the Voidoids. 1977. *Blank Generation.* Rhino.
Roach, Steve, Kevin Braheny and Michael Stearns. 1989. *Desert Solitaire.* Fortuna.
Rollins, Sonny. 1998. *Global Warming.* Milestone.
Ronstadt, Linda. 1969. *Hand Sown ... Home Grown.* EMI.
Rothenberg, David. 2002. *Sudden Music.* (CD with book).
Rovics, David. 2004. *Songs for Mahmud.* Ever Reviled Records. On line at: david.rovics.com
Sanders, Pharoah. 1969. *Karma.* Impulse!
Saunders, Merl. 1990. *Blues from the Rainforest: A Musical Suite.* Sumertone.
Scott-Heron, Gil and Brian Jackson. 2005. *From South Africa to South Carolina.* TvT.
Seeger, Pete. 1965. *Strangers and Cousins.* Columbia.
– 1966. *God Bless The Grass.* Columbia/Legacy.
– 1967. *Pete Seeger's Greatest Hits.* Columbia/Legacy.
– 1968. *Waist Deep In The Big Muddy.* Columbia/Legacy.
– 2000. 'Mrs Clara Sullivan's Letter' on Various Artists. *The Best of Broadside 1962-1988: Anthems of the American Underground from the Pages of Broadside Magazine.* Smithsonian Folkways.
Shamen, The. 1999. *Boss Drum.* One Little Indian.
Six Organs of Admittance. 1998. *Dust and Chimes.* Holy Mountain.
Smith, Harry. 1998. *Anthology of American Folk Music.* Smithsonian Folkways.
Smith, Patti. *Horses.* Arista. 2005.
Spirit. 1970. *Twelve Dreams of Dr. Sardonicus.* Sony/Legacy.
Steppenwolf. 2000. *Born to be Wild.* Gfs.
Stoltz, Walkin' Jim. 1984. *Spirit Is Still On The Run.* Wild Wind Music.
– 1989. *Listen To The Earth.* Wild Wind Music.
Sun Ra. 1958. *Jazz in Silhouette.* Evidence.
– 1972. *Space Is the Place.* Blue Thumb.
Talking Heads. 2006. *Naked.* Rhino/WEA.
Taylor, James. 1999. *One Man Dog.* Warner.
Ten Years After. 1971. *The Essential Ten Years After Collection.* Chrysalis.
Various Artists. 1974. *Clearwater.* Hudson River Sloop Restoration, Inc.
– 1999a. *Fish Tree Water Blues.* Bullseye Blues and Jazz.
– 1999b. *No Nukes.* Warners.
Welch, Gillian. 1996. *Revival.* Acony.
Winter, Paul and Paul Halley. 1987. *Whales Alive.* Living Music.
Wonder, Stevie. 1972. *Innervisions.* Motown.
– 1979. *Journey Through the Secret Life of Plants.* Motown.
Young, Adrienne and Little Sadie. 2004. *Plow to the End of the Row.* AddieBelle Music.
Young, Neil. 1970. *After the Goldrush.* Reprise.
– 1977. *American Stars and Bars.* Warners.
– 1983. *Old Ways.* Geffen.
Young, Neil and Crazy Horse. 2004. *Greendale.* Reprise.
Zappa, Frank. 1993. *The Yellow Shark.* Rykodisc.
Zappa/Beefheart/Mothers. 1975. *Bongo Fury.* Rykodisc.
Zappa, Frank and the Mothers of Invention. 1966. *Freak Out!* Rykodisc.
– 1968. *We're Only In It For The Money.* Rykodisc.

– 1969. *Uncle Meat.* Rykodisc.
– 1970. *Weasels Ripped My Flesh.* Rykodisc.
– 1971. *200 Motels.* Rykodisc.
– 1972. *Waka/Jawaka.* Rykodisc.
– 1993. *Ahead of Their Time.* Rykodisc.

# Works Cited


Abram, David. 1997. *The Spell of the Sensuous: Perception and Language in a More-Than-Human World.* New York: Vintage Books.

Adams, Henry. 1961. *The Education of Henry Adams: An Autobiography.* Boston: Houghton and Mifflin.

Adorno, Theodor W. 1948. *Philosophy of Modern Music* (tr. A. G. Mitchell and W. V. Bloomster). London: Sheed and Ward.

– 1974. *Minima Moralia: Reflections From A Damaged Life* (tr. E. F. N. Jephcott). London: New Left Books.

– 1984. *Aesthetic Theory* (tr. C. Lenhardt). London: Routledge and Kegan Paul.

– 1988. *Introduction to the Sociology of Music* (tr. E. B. Ashton). New York: Continuum.

– 2002. *Essays on Music* (tr. S. H. Gillespie). Leppert, Richard (ed.). Berkeley, Los Angeles and London: University of California Press.

Albanese, Catherine L. 1990. *Nature Religion in America: From the Algonkian Indians to the New Age.* Chicago and London: University of Chicago Press.

Alperson, Philip (ed.). 1987. *What is Music? An Introduction to the Philosophy of Music.* University Park, Pennsylvania: Pennsylvania State University Press.

Anderson, Terry H. 1995. *The Movement and the Sixties.* New York and Oxford: Oxford University Press.

Athanasiou, Tom. 1997. *Slow Reckoning: The Ecology of a Divided Planet.* London: Secker and Warburg.

Attali, Jacques. 1985. *Noise: The Political Economy of Music* (tr. B. Massumi). Minneapolis: University of Minneapolis Press.

Baker, Houston A. 1984. *Blues, Ideology and Afro-American Literature: A Vernacular Theory.* Chicago and London: University of Chicago Press.

Bamberger, W. C. 1999. *Riding Some Kind of Unusual Skull Sleigh: On the Arts of Don Van Vliet.* Whitmore Lake, Michigan: Alap Editions.

Bangs, Lester. 1987. *Psychotic Reactions and Carburetor Dung.* London: Serpent's Tail.

Banks, Jack. 1996. *Monopoly Television: MTV's Quest to Control the Music.* Oxford: Westview Press.

Barnes, Mike. 2000. *Captain Beefheart.* London: Quartet Books.

Bataille, Georges. 2006. 2006. *Eroticism.* London: Marion Boyars.

Bate, Jonathan. 2000. *The Song of the Earth.* London: Picador.

Bateson, Gregory. 1972. *Steps to An Ecology Of Mind: A Revolutionary Approach To Man's Understanding of Himself.* New York: Ballantine Books.

Becker, Judith. 2001. 'Anthropological Perspectives on Music and Emotion' in Juslin and Sloboda (eds). (2001): 135-160.

Benjamin, Walter. 1973. *Illuminations.* Fontana: Glasgow.

Bennett, Michael. 2001. 'Anti-Pastoralism, Frederick Douglass, and the Nature of Slavery' in Armbruster, Karla and Kathleen R. Wallace (eds). *Beyond Nature*

*Writing: Expanding the Boundaries of Ecocriticism*. Charlottesville and London:University Press of Virginia: 195-201.

Bennett, Michael and Teague, David W. (eds). 1999. *The Nature of Cities: Ecocriticism and Urban Environments*. Tucson: University of Arizona Press.

Berendt, Joachim-Ernst. 1988. *Nada Brahma: The World Is Sound: Music and the Landscape of Consciousness* (tr. H. Bredigkeit). London and the Hague: East West Publications.

Berke, Joseph (ed.). 1969.*Counter Culture*. London: Owen.

Berliner, Paul. 1994. *Thinking in Jazz: The Infinite Art of Improvisation*. Chicago and London: University of Chicago Press.

Berman, Marshall. 1982. *All That Is Solid Melts Into Air: The Experience of Modernity*. London: Verso.

Berman, Morris. 1981. *The Reenchantment of the World*. Ithica and London: Cornell University Press.

Berry, Wendell. 1977. *The Unsettling of America: Culture and Agriculture*. San Francisco: Sierra Club Books.

*Better Living Through Circuitry*. Dir. Jon Reiss. Cleopatra Pictures. USA. 1999. (film).

Betts, Kellyn S. 1994. 'The Coming Green Computers' in *E! The Environmental Magazine* 5.2 (March/April 1994): 28-35.

Biehl, Janet. 1991. *Rethinking Ecofeminist Politics*. Boston: South End Press.

Bloch, Ernst. 1995. *The Principle of Hope*. Volumes I-III (tr. N. Plaice, S. Plaice and P. Knight). Cambridge, Massachusetts: MIT Press.

Bohlman, Philip V. 2002. *World Music: A Very Short Introduction*. Oxford: Oxford University Press.

Boon, Marcus. (nd). 'Jon Hassell: There Was No Avant Garde'. On line at: http://www.hungryghost.net/PPN/Hassell.htm (consulted 29.03.2006).

Boone, Graeme M. 1997. 'Tonal and Expressive Ambiguity in Dark Star' in Covach, John and Graeme M. Boone (eds). *Understanding Rock: Essays in Musical Analysis*. New York and Oxford: Oxford University Press, 171-209.

Born, Georgina and David Hesmondhalgh (eds). 2000. *Western Music and Its Others: Difference, Representation and Appropriation in Music*. Berkeley and London: University of California Press.

Botkin, Daniel B. 1990. *Discordant Harmonies: A New Ecology for the Twenty-First Century*. New York and Oxford: Oxford University Press.

Boucher, Caroline. 1972. 'Zappa Stole My Ideas, Says the Captain'. On line at: http://www.beefheart.com (consulted 21.08.2006).

Boyce-Tillman, June. 2000. *Constructing Musical Healing: The Wounds That Heal*. London and Philadelphia: Jessica Kingsley Publishers.

Boyes, Georgina. 1993. *The Imagined Village: Culture, Ideology and the English Folk Revival*. Manchester and New York: Manchester University Press.

Bramwell, Anna. 1989. *Ecology in the 20th Century: A History*. New Haven and London: Yale University Press.

Brand, Stewart. 1999. *The Clock of the Long Now: Time and Responsibility*. London: Weidenfeld and Nicolson.

Braun, Hans-Joachim. 2002. 'Moving On: Trains and Planes as a Theme in Music' in Braun, Hans-Joachim (ed.) *Music and Technology in the Twentieth Century*. Baltimore and London: Johns Hopkins Press: 106-20.

Buck-Morss, Susan. 1977. *The Origin of Negative Dialectics: Theodor W. Adorno, Walter Benjamin,and the Frankfurt Institute.* Hassocks: The Harvester Press.

Buchanan, Ian and Marcel Swiboda (eds). 2004. *Deleuze and Music.* Edinburgh: Edinburgh University Press.

Budiansky, Stephen. 1996. *Nature's Keepers: The New Science of Nature Management.* London: Phoenix Giant,

Buell, Lawrence. 1995. *The Environmental Imagination: Thoreau, Nature Writing, and the Formation of American Culture.* Cambridge, Massachusetts and London: The Belknap Press of Harvard University.

– 2001. *Writing for an Endangered World: Literature, Culture, and Environment in the U.S. and Beyond.* Cambridge, Massachusetts and London: The Belknap Press of Harvard University.

Buxton, David. 1990. 'Rock Music, the Star System, and the Rise of Consumerism' in Frith, Simon and Andrew Goodwin. *On Record: Rock, Pop, and the Written Word.* London: Routledge.

Cage, John. 1961. *Silence.* Cambridge, Massachusetts and London: MIT Press.

– 1968. *A Year From Monday.* London: Calder and Boyars.

– 1981. In conversation with Daniel Charles. *For The Birds.* Boston and London: Marion Boyars.

Callicott, J. Baird and Roger T. Ames (eds). 1989. *Nature in Asian Traditions of Thought: Essays in Environmental Philosophy.* Albany: State University of New York Press.

Campbell, Gregg M. 1998. 'Bob Dylan and the Pastoral Apocalypse' in Benson, Carl (ed.) *The Bob Dylan Companion: Four Decades of Commentary.* New York: Schirmer Books: 97-108.

Cantwell, Robert. 1996. *When We Were Good: The Folk Revival.* Cambridge, Massachusetts and London: Harvard University Press.

Capra, Fritjof. 1975. *The Tao of Physics.* London: Wildwood House.

– 1982. *The Turning Point: Science, Society and the Rising Culture.* London: Flamingo.

Carlin, Richard. 1995. *The Big Book of Country Music: A Biographical Encyclopedia.* Harmondsworth: Penguin.

Carpenter, Edmund and Marshall McLuhan (eds). 1960. *Explorations in Communication.* Boston: Beacon Press.

Carriker, Robert. C. 2001. 'Ten Dollars a Song: Woody Guthrie Sells His Talent to the Bonneville Power Administration'. On line at: http://www.wshs.org/wshs/Columbia/articles/0101-a2.htm (consulted 14.08.2006).

Carson, Rachel. 1991. *Silent Spring.* Harmondsworth: Penguin.

Center for Black Music Research. (nd). 'Country Music: Definition of Style'. On line at: http://www.cbmr.org/styles/country.htm (consulted 23.08.2006).

Chanan, Michael. 1994. *Musica Practica: The Social Practice of Western Music from Gregorian Chant to Postmodernism.* London and New York: Verso.

Chapman, Mary and Glenn Hendler (eds). 1999. *Sentimental Men: Masculinity and the Politics of Affect in American Culture.* Berkeley, Los Angeles, London: University of California Press.

Charlesworth, Chris (ed.). 1993. *Frank Zappa in His Own Words.* London: Omnibus Press.

Chorush, Bob. 1971. 'Captain Beefheart's Maidenform Rock.' On line at: http://www.beefheart.com (consulted 11.08.2006)

Clarke, J. J. 2000. *The Tao of the West: Western Transformations of Taoist Thought.* London and New York: Routledge.

Clayton, Martin, Rebecca Sager and Udo Will. 2005. 'In Time With the Music: the Concept of Entrainment and its Significance for Ethnomusicology' in *European Meetings in Ethnomusicology* 11: 3-75.

Cold, Jonathan. 1994. 'Sex, Pain and Rock & Roll'. *Rolling Stone* (8 September 1994): 51-4.

Collett-White, Mike. 2007. 'Earth Underwhelmed by Environment Pop Extravaganza'. On line at: http://www/alertnet.org/thenews/newsdesk/L08203765.htm (consulted 7.04.2009).

Commoner, Barry. 1971. *The Closing Circle: Nature, Man and Technology.* New York: Alfred A. Knopf.

Corbett, John. 2000. 'Experimental Oriental: New Music and Other Others' in Born, Georgina and David Hesmondhalgh (eds). (2000): 163-86.

Courrier, Kevin. 2002. *Dangerous Kitchen: The Subversive World of Frank Zappa.* Toronto: ECW Press.

Cowley, Julian. 2005. 'The Life Aquatic'. *The Wire.* 255 (May 2005): 26-27.

Cox, Christoph and Daniel Warner (eds). 2004. *Audio Culture: Readings in Modern Music.* New York and London: Continuum.

Cronon, William. 1996. 'The Trouble with Wilderness' in Cronon, William (ed.) *Uncommon Ground: Rethinking the Human Place in Nature.* New York and London: W. W. Norton: 69-90.

Crosby, David and Gottlieb, Carl. 1989. *Long Time Gone: The Autobiography of David Crosby.* London: Heinemann.

Davis, Erik. 1998. *Techgnosis: Myth, Magic and Mysticism in The Age Of Information.* London: Serpent's Tail.

Davis, Mike. 1990. *City of Quartz: Excavating the Future in Los Angeles.* London:Vintage.

Davis, R. G. 2002. 'Music from the Left'. On line at: http://www.usopera-web.com/2002/jan/left.htm (consulted 10.10.2004).

Davis, Sharon. 1991. *I Heard It Through The Grapevine. Marvin Gaye: The Biography.* Edinburgh: Mainstream Publishing.

De Jonge, Eccy. 2004. *Spinoza and Deep Ecology: Challenging Traditional Approaches to Environmentalism.* Aldershot: Ashgate.

Deleuze, Gilles. 1983. *Nietzsche and Philosophy* (tr. H. Tomlinson). London: Athlone Press.

Deleuze, Gilles, and Félix Guattari. 1987. *A Thousand Plateaus: Capitalism and Schizophrenia* (tr. B. Massumi). London: Athlone Press.

– 1994. *What Is Philosophy?* (tr. H. Tomlinson and G. Burchell). New York: Columbia University Press.

De Man, Paul. 1983. *Blindness and Insight: Essays in the Rhetoric of Contemporary Criticism.* London: Methuen.

DeMott, Benjamin. 1970. 'Rock as Salvation' in White, David Manning (ed.) *Popular Culture in America.* Chicago: Quadrangle Books: 191-204.

Denisoff, R. Serge. 1971. *Great Day Coming: Folk Music and the American Left.* Urbana: University of Illinois Press.

Denselow, Robin. 1989. *When the Music's Over: The Story of Political Pop.* London and Boston: Faber and Faber.

Dobra, Susan. 2002. 'Alice di Micele: Singing Out about the Beauty of the Earth' in *Magical Blend* (July 2002): 54-7.

Dobson, Andrew. 1993. 'Critical Theory and Green Politics' in Dobson, Andrew and Paul Lucardie (eds). *The Politics of Nature: Explorations in Green Political Theory*. London and New York: Routledge: 190-209.

– 1998. *Justice and the Environment: Conceptions of Environmental Sustainability and Dimensions of Social Justice*. Oxford and New York: Oxford University Press.

– 2000. *Green Political Thought*. London and New York: Routledge.

Doggett, Peter. 2000. *Are You Ready For the Country: Elvis, Dylan, Parsons and the Roots of Country Rock*. London: Penguin.

Downing, David. 1976. *Future Rock*. St. Albans: Panther Books.

Dunaway, David King. 1985. *How Can I Keep From Singing: Pete Seeger*. London: Harrap.

Dunn, David. 1988. 'Wilderness as Reentrant Form: Thoughts on the Future of Electronic Art'. On-line at: http://www.daviddunn.com/~david/writings/wilder.pdf (consulted 22.07.09).

Durant, Alan. 1989. 'Improvisation in the Political Economy of Music' in Norris, Christopher (ed.) *Music and the Politics of Culture*. London: Lawrence and Wishart: 252-282.

Dyer, Richard. 1985. 'Entertainment and Utopia' in Nichols, Bill (ed.) *Movies and Methods vol. II*. Berkeley: University of California Press.

Eckersley, Robyn. 1992. *Environmentalism and Political Theory: Toward an Ecocentric Approach*. London: UCL Press.

Ehrlich, Paul R. 1971. *The Population Bomb*. London: Pan Books.

Eisenberg, Evan. 1987. *The Recording Angel: Music, Records and Culture from Aristotle to Zappa*. London: Picador.

– 1998. *The Ecology of Eden*. New York: Vintage.

Eisler, Hanns and Adorno, Theodor. 2004. 'The Politics of Hearing' in Cox and Warner (eds.) (2004): 73-5.

Emerson, Ralph Waldo. 1894. *Society and Solitude*. Boston: Houghton, Mifflin and Co.

England, Phil. 2002. 'Making the Nature Scene' in *The Wire*. 226 (December 2002): 40-7.

Eno, Brian. 1996. *A Year with Swollen Appendices*. London and Boston: Faber and Faber.

European Jazz Network. (nd). 'Sonny Rollins: tenor sax'. On-line at: http://www.ejn.it/ mus /rollins.htm (consulted 14.09.2004).

Evans, Mei Mei. 2002. '"Nature" and Environmental Justice' in Adamson, Joni, Mei Mei Evans and Rachel Stein (eds). *The Environmental Justice Reader: Politics, Poetics, & Pedagogy*. Tucson: University of Arizona Press: 181-93.

Evernden, Neil. 1985. *The Natural Alien*. Toronto, Buffalo and London: University of Toronto Press.

Fasteau, Kali Z. (nd). 'Statement of Purpose'. On line at: http:// www.kalimuse.com (consulted 18.14.2006).

Feld, Steven. 1994. 'Aesthetics as Iconicity of Style (uptown title); or, (downtown title) "Lift-up-over-Sounding": Getting into the Kaluli Groove' in Keil, Charles and Steven Feld (eds). *Music Grooves: Essays and Dialogues*. Chicago and London: University of Chicago Press: 109-50.

Fink, Robert. 2005. *Repeating Ourselves: American Minimal Music as Cultural Practice*. Berkeley and London: University of California Press.

Fisher, John Andrew. 1998. 'What the Hills Are Alive With: In Defense of the Sounds of Nature' in *The Journal of Aesthetics and Art Criticism* 56.2 (Spring 1998): 167-179.

Fiske, John. 1989. *Understanding Popular Culture* Boston: Unwin Hyman.

Fong-Torres, Ben. 1970.'The Rolling Stone Interview: Grace Slick with Paul Kantner'. *Rolling Stone* (12 November 1970). On line at: http://www.math.ufl.edu/~jaz/Articles/paulgrace (consulted 6.02.2004).

Forsyth, Tim. 2001. 'Critical Realism and Political Ecology' in López, José and Gary Potter (eds). *After Postmodernism: An Introduction to Critical Realism*. London: Continuum: 146-154.

Fox, Warren. 1990. *Toward a Transpersonal Ecology: Developing New Foundations for Environmentalism*. Boston and London: Shambhala.

Frame, Pete. 1969. 'A Fine Madness'. On line at: http://www.beefheart.com (consulted 18.08.2006).

Frank, Thomas. 1997. *The Conquest of Cool: Business Culture, Counterculture, and the Rise of Hip Consumerism*. Chicago and London: University of Oregon Press.

*Free Willy*. 1993. Dir. Simon Wincer. Warner Bros. USA. (film).

Freud, Sigmund. 1963. *Civilization and Its Discontents* (tr. J. Rivière). Strachey, James (ed.). London: Hogarth Press.

Frith, Simon. 1984. 'Rock and the Politics of Memory' in Sayres, Sohnya, Anders Stephanson, Stanley Aronowitz and Frederic Jameson (eds). *The 60s Without Apology*. Minneapolis: University of Minneapolis Press in cooperation with *Social Text*.

– 2000. 'The Discourse of World Music' in Born, Georgina and David Hesmondhalgh (eds). Berkeley and London: University of California Press: 305-22.

Fubini, Enrico. 1987. *The History of Music Aesthetics* (tr. M. Hatwell). Basingstoke and London: Macmillan.

Fuller, R. Buckminster. 1969. *Operating Manual for Spaceship Earth*. Carbondale: S. Illinois University Press.

Fusion Anomaly. (nd). On-line at: http:www.fusionanomaly.net/newedge.html (consulted 06.06.2006)

Gablik, Suzi. 1991. *The Reenchantment of Art*. New York: Thames and Hudson.

Gans, Herbert. 1974. *Popular Culture and High Culture: An Analysis and Evaluation of Taste*. New York: Basic Books.

Gardner, Kay. 1990. *Sounding the Inner Landscape: Music as Medicine*. Rockport, Massachusetts, Shaftesbury, Dorset, Brisbane, Queensland: Element.

Garrard, Greg. 2004. *Ecocriticism*. Abingdon: Routledge.

Gaut, Berys. 2001. 'Art and Ethics' in Gaut, Berys and Dominic McIver (eds). *The Routledge Companion to Aesthetics*. London and New York: Routledge: 341-52.

Gendron, Bernard. 1986. 'Theodor Adorno Meets the Cadillacs' in Modleski, Tania (ed.) *Studies in Entertainment: Critical Approaches to Mass Culture*. Bloomington and Indianapolis: Indiana University Press.

Geoghegan, Vincent. 1996. *Ernst Bloch*. London and New York: Routledge.

Gifford, Terry. 1999. *Pastoral*. London and New York: Routledge.

Ginsberg, Allen. 1969. 'Consciousness and Practical Action' in Berke, Joseph (ed.).(1969): 170-81

Gitlin, Todd. 1987. *The Sixties: Years of Hope, Days of Rage*. New York: Bantam Books.
Godwin, Joscelyn. 1987. *Harmonies of Heaven and Earth: The Spiritual Dimension of Music from Antiquity to the Avant-Garde*. London: Thames and Hudson.
Gould, Stephen J. 1990. *An Urchin in the Storm*. Harmondsworth: Penguin.
Gracyk, Theodore. 1996. *Rhythm and Noise: An Aesthetics of Rock*. London and New York: I. B. Tauris & Co Ltd.
Graves, John. 1960. *Goodbye to a River: A Narrative*. New York: Knopf.
Greenway, John. 1960. *American Folksongs of Protest*. New York: A. S. Barnes.
Griffiths, Paul. 1995. *Modern Music and After: Directions since 1945*. Oxford: Oxford University
Gross, Paul R. and Norman Levitt. 1994. *Higher Superstition: The Academic Left and Its Quarrels with Science*. Baltimore: Johns Hopkins University Press.
Gross, Paul R., Norman Levitt and Martin W. Lewis (eds). 1996. *The Flight from Science and Reason*. New York: New York Academy of Sciences.
Grundmann, Reiner. 1991. *Marxism and Ecology*. Oxford: Clarendon Press.
Habermas, Jürgen. 1990. *The Philosophical Discourse of Modernity: Twelve Lectures*. Cambridge, Massachusetts : MIT Press.
Hainge, Greg. 2004. 'Is Pop Music?' in Buchanan, Ian and Marcel Swiboda, Marcel (eds). *Deleuze and Music*. Edinburgh: Edinburgh University Press: 36-53.
Halpern, Steven. 1998. 'Introduction' in Werkhoven, Henk N.(ed.) *The International Guide to New Age Music*. New York: Billboard Books: viii-ix.
Hanfling, Oswald. 2001. 'Wittgenstein on Language, Art and Humanity' in Allen, Richard and Malcolm Turvey (eds). *Wittgenstein, Theory and the Arts*. London and New York: Routledge: 75-91.
Haraway, Donna J. 1991. *Simians, Cyborgs, and Women: The Reinvention of Nature*. New York: Routledge.
Harron, Mary.1988. 'McRock: Pop as a Commodity', in Frith, Simon (ed.) *Facing the Music: Essays on Pop, Rock and Culture*. New York: Mandarin: 173-220.
Hart, Mickey and Lieberman, Fredric. 1991. *Planet Drum*. San Francisco: Harper San Francisco.
Harvey, David. 1989. *The Condition of Postmodernity: An Enquiry into the Origins of Cultural Change*. Oxford: Basil Blackwell.
– 1996. *Justice, Nature and the Geography of Difference*. Oxford, UK and Cambridge, Massachusetts: Blackwell.
Hassell, Jon. 1982. 'Artificial boundaries, expanding horizons, possible musics'. On line at: http://www.jonhassell.com/heavy.html (consulted 29.03.2006).
Hawken, Paul. 1993. *The Ecology of Commerce: A Declaration of Sustainability*. New York: HarperCollins.
Haynes, Jim. 2001. 'John Hudak: Time Passes Slowly'. *The Wire*. 209 (July 2001): 14.
Heinlein, Robert. 1980. *Stranger in a Strange Land*. London: New English Library.
Heise, Ursula K. 1997. 'Science and Ecocriticism'. On line at: http://www.asle.org/site/-resources/ecocritical-library/intro/science (consulted 04.02.2009).
Henley, Don and Dave Marsh (eds). 1991. *Heaven Is Under Our Feet: A Book for Walden Woods*. New York: Berkley Books.
Hesmondhalgh, David. 2000. 'International Times: Fusions, Exoticism and Antiracism in Electronic Dance Music' in Born, Georgina and David Hesmondhalgh (eds). (2000): 280-304.

Heylin, Clinton. 2000. *Bob Dylan Behind the Shades: The Biography Take Two*. London: Viking.

Hollings, Ken. 1998. 'Apocalypse, Hawaiian Style'. *The Wire*. 174 (August 1998): 40-5.

Horkheimer, Max and Theodor Adorno. 1973. *Dialectic of Enlightenment* (tr. J. Cumming). London: Allen Lane.

Hoskyns, Barney. 1996. *Waiting for the Sun: The Story of the Los Angeles Music Scene*. London: Viking.

Hutnyk, John. 2000. *Critique of Exotica: Music, Politics and the Culture Industry*. London and Sterling, VA: Pluto Press.

Hutson, Scott R.1999. 'Technoshamanism: Spiritual Healing in the Rave Subculture' in *Popular Music and Society*. 23.3 (Fall 1999): 53-77.

Ingram, David. 2006. 'A balance that you can hear: Ecology and the Soundscape Recordings of David Dunn' in *European Journal of American Culture* 25 (2): 123-38.

– 2007(a). 'The clutter of the unkempt forest: Ecology and the Music of John Cage' in *Amerikastudien/ American Studies* 51 (4): 565-79.

– 2007(b). 'For Free?: Theorising Consumption, Commerce, and the Environmental Costs of Artistic Production'. *Green Letters* 8: 13-22.

Jameson, Fredric. 1991. *Postmodernism, or; the Cultural Logic of Late Capitalism*. London and New York: Verso.

Jay, Martin. 1993. *Downcast Eyes: The Denigration of Vision in Twentieth-Century French Thought*. Berkeley: University of California Press.

John, Eileen. 2001. 'Art and Knowledge' in Gaut, Berys and Dominic McIver (eds). *The Routledge Companion to Aesthetics*. London and New York: Routledge: 329-40.

Jones, Le Roi. 1963. *Blues People: Negro Music in White America*. New York: William Morrow.

Juslin, Patrik N. and John A Sloboda (eds). 2001. *Music and Emotion: Theory and Research*. Oxford: Oxford University Press.

Kahn, Douglas. 1999. *Noise, Water, Meat: A History of Sound in the Arts*. Cambridge, Massachusetts and London: MIT Press.

Keil, Charles. 1966. *Urban Blues*. Chicago and London: University of Chicago Press.

Keil, Charles and Patricia Campbell. 2006. *Born to Groove*. On line at http://www.borntogroove.com (consulted 18.08.2008).

Keil, Charles and Steven Feld (eds). *Music Grooves: Essays and Dialogues*. Chicago and London: University of Chicago Press.

Kidner, David W. 2000. *Nature and Psyche: Radical Environmentalism and the Politics of Subjectivity*. Albany, NY: State University of New York Press.

Kirkman, Robert. 2002. *Skeptical Environmentalism: The Limits of Philosophy and Science*. Bloomington and Indianapolis: Indiana University Press.

Kivy, Peter. 1984. *Sound and Semblance: Reflections on Musical Representation*. Princeton: Princeton University Press.

*Koyaanisqatsi*. Dir. Godfrey Reggio. Island Alive. USA, 1983. (film).

Krause, Bernie.1998. *Into A Wild Sanctuary: A Life in Music and Natural Sound*. Berkeley, California: Heyday Books.

Krech III, Shepard. 1999. *The Ecological Indian: Myth and History*. New York and London: W.W. Norton.

Krims, Adam. 2000. *Rap Music and the Poetics of Identity*. Cambridge: Cambridge University Press.

Kristeva, Julia. 1984. *Revolution in Poetic Language*. New York: Columbia Umiversity Press.

Kuklick, Bruce. 2001. *A History of Philosophy in America*. Oxford: Clarendon Press.

Kupferberg, Tuli. 1969. 'The Coming Catastrophic Age of Leisure' in Berke, Joseph (ed.) (1969): 74-89.

Landau, Jon. 1968. 'John Wesley Harding' in McGregor, Craig (ed.) 1980: 108-20.

Lange, Art. 1998. 'The Primer: Musique Concrète'. *The Wire* 174 (August 1998): 50-55.

Lanza, Joseph. 2004. *Elevator Music: A Surreal History of Muzak, Easy-Listening, and Other Moodsong*. Ann Arbor: University of Michigan Press.

Lawley, Guy. 1999. 'I Like Hate and I Hate Everything Else: the Influence of Punk on Comics' in Sabin (ed.) (1999): 100-19.

Leary, Timothy. 1968. *The Politics of Ecstasy*. Berkeley: Ronin.

Lee, Martha F. 1995. *Earth First! Environmental Apocalypse*. Syracuse, NY: Syracuse University Press.

Lee, Martin A. and Shlain, Bruce. 1985. *Acid Dreams: The Complete Social History of LSD: The CIA, the Sixties, and Beyond*. New York: Grove Press.

Lefèbvre, Henri. 1991. *The Production of Space* (tr. D. Nicholson-Smith). Oxford, UK and Cambridge, Massachusetts: Blackwell.

Leopold, Aldo. 1948. *A Sand County Almanac, and Sketches Here and There*. London: University Press.

Lester, Julius. 1966. 'The Angry Children of Malcolm X'. *Sing Out!* 16.5 (October – November 1966): 21-5.

Levinson, Paul. 1999. *Digital McLuhan: A Guide to the Information Millennium*. London and New York: Routledge.

Levitas, Ruth. 1990. *The Concept of Utopia*. London: Philip Allan.

Lewis, George E. 2004. 'Improvised Music after 1950: Afrological and Eurological Perspectives' in Cox and Warner (eds). (2004): 272-284.

Lewis, Martyn W. 1992. *Green Delusions: An Environmentalist Critique of Radical Environmentalism*. Durham and London: Duke University Press.

Lieberman, Robbie. 1989. *'My Song Is My Weapon': People's Songs, American Communism, and the Politics of Culture, 1930-1950*. Urbana and Chicago: University of Illinois Press.

Light, Andrew. 1999. 'Boyz in the Woods: Urban Wilderness in American Cinema' in Bennett, Michael and David W. Teague (eds). (1999): 137-156.

Limerick, Patricia. 1987. *The Legacy of Conquest: The Unbroken Past of the American West*. New York and London: Norton.

Lock, Graham. 1999. *Blutopia: Visions of the Future and Revisions of the Past in the Work of Sun Ra, Duke Ellington, and Anthony Braxton*. Duke University Press: Durham and London.

Lomax, John A. 1916. *Cowboy Songs and Other Frontier Ballads*. New York: Sturgis and Walton.

Lone Wolf Circles. 1985a. 'The Deep Ecology Soundtrack: Part 2: Let the Animals Sing'. *Earth First!* (1 August 1 1985): 25.

– 1985b. 'The Country in Country Music'. *Earth First!* 22 September 1985: 28.

– 1988. 'Part XIII: Amerindian Riffs.' *Earth First!* 2 February 2 1988: 24.

– 1991. *Full Circle: A Song of Ecology and Earth Spirituality*. Llewellyn: St. Paul MN.

Lovelock, James. 2004. 'Nuclear Power is the only green solution.' *The Independent* (24 May 2004). On line at: http://www/ecolo.org/media/articles/articles.in.english/love-indep-24-05-04.htm (consulted 12.05.09).
Lukács, Georg. 1971. *History and Class Consciousness: Studies in Marxist Dialectics* (tr. R. Livingstone). London: Merlin Press.
Lydon, Michael. 1972. 'The Grateful Dead' in La Valley, Albert J. (ed.) *The New Consciousness*. Cambridge, Massachusetts: Winthrop.
McClary, Susan. 1991. *Feminine Endings: Music, Gender, and Sexuality*. Minnesota and London: University of Minnesota Press.
– 2000. *Conventional Wisdom: The Content of Musical Form*. Berkeley, Los Angeles and London: University of California Press.
McCulloch, Alistair. 2007. 'Live Earth and the Green Paradox'. On line at: http://pop-music.suite101.com/article.cfm/live earth and the green paradox (consulted 7.04.2009).
McGrath, Rick. 1971. 'A Candid Conversation with Beef The Chief'. On line at: http://www.beefheart.com (consulted 16.07.06)
– 1973. 'Know What I Mean? The Don Van Vliet Interview'. On line at: http://www.beefheart.com (consulted 16.07.06).
McGregor, Craig (ed.) 1980. *Bob Dylan: A Retrospective*. London: Angus & Robertson.
McKenna, Terence. 1992. *The Archaic Revival*. San Francisco: HarperSanFrancisco.
McKnight, Connor. 1972. 'Jottings for the Beefheart Archives'. *Zig Zag* (29 March 1972). On line at: http://www.beefheart.com (consulted 16.07.06).
McLuhan, Marshall. 1962. *The Gutenberg Galaxy: The Making of Typographic Man*. London: Routledge & Kegan Paul.
– 1967. *Understanding Media: The Extensions of Man*. London: Sphere.
– 2004. 'Visual and Acoustic Space' in Cox, Christoph and Warner, Daniel (eds). (2004): 67-75.
McLuhan, Marshall and Quentin Fiore 1967. *The Medium is the Massage: an Inventory of Effects*. Harmondsworth: Penguin.
McLuhan, Marshall with Wilfred Watson 1971. *From Cliché to Archetype*. New York: Pocket Books.
McPherson, C. B. 1962. *The Political Theory of Possessive Individualism: Hobbes to Locke*. Oxford: Oxford University Press.
MacKenzie, John M. 1995. *Orientalism: History, Theory and the Arts*. Manchester: Manchester University Press.
Mailer, Norman. 1970. *A Fire on the Moon*. London: Weidenfeld and Nicolson.
Maltman, Phil. 1972. 'Beefheart Interview'. On line at: http://www.freewebs.com/tee-jo/argue/drive.html (consulted 23.08.06).
Marcus, Greil. 1997 *Invisible Republic: Bob Dylan's Basement Tapes*. New York: Picador.
Marx, Karl. 1964. *The Economic and Philosophical Manuscripts of 1844* (tr. M. Milligan). Dirk J. Struik (ed.) New York: International Publishers.
Marx, Leo. 1964. *The Machine in the Garden: Technology and the Pastoral Ideal in America*. Oxford: Oxford University Press.
Matthews, Freya. 1999. *The Ecological Self*. London: Routledge.
Meadows, Donella H., Dennis L. Meadows, Jorgen Randers, William W. Behrens III. 1972. *The Limits to Growth: a Report for the Club of Rome's Project on the Predicament of Mankind*. London: Earth Island.

Mellers, Wilfred. 1984. *A Darker Shade of Pale: A Backdrop to Bob Dylan*. London: Faber and Faber.
Merchant, Carolyn. 1982. *The Death of Nature: Women, Ecology, and the Scientific Revolution*. London: Wildwood House.
Meyer, Leonard B. 1967. *Music, the Arts and Ideas*. London: University of Chicago.
Miller, Timothy. 1991. *The Hippies and American Values*. Knoxville: University of Tennessee Press.
Moore, Allan F. 2001. *Rock: The Primary Text*. Aldershot: Ashgate.
Morse, Jim and Nancy Mathews (eds). 1971. *Survival Songbook*. San Francisco and New York: The Sierra Club.
Morton, Timothy. 2007. *Ecology without Nature: Rethinking Environmental Aesthetics*. Cambridge, Massachusetts and London: Harvard University Press.
Mottram, Eric. 1988. *Blood on the Nash Ambassador*. London: Hutchinson Radius.
Mumford, Lewis. 1970. *The Myth of the Machine: The Pentagon of Power*. New York: Harcourt Brace Jovanovich.
Murphie, Andrew. 2003. 'Electronicas: Differential Media and Proliferating, Transient Worlds'. On line at: http://www.fineartforum.org/Backissues/Vol 17/faf v17 n08/reviews/Murphie.html (consulted 25.11.04).
Murphy, Timothy S. and Daniel W. Smith. 2001. 'What I Hear Is Thinking Too: Deleuze and Guattari Go Pop'. *Echo: A Music-Centered Journal* 3.1 (Spring 2001). On line at: http://www.echo.ucla.edu/Volume3-Issue1
Murray, Charles Shaar. 1989. *Crosstown Traffic: Jimi Hendrix and-Post-war Pop*. London and Boston: Faber and Faber.
Naess, Arne. 1989. *Ecology, Community and Lifestyle* (tr. D. Rothenberg). Cambridge: Cambridge University Press.
Nelson, Paul. 1965. 'Newport Folk Festival, 1965' in McGregor, Craig (ed.) (1980): 30-5.
*New Jack City*. Dir. Mario van Peebles. Warner Bros. USA, 1991. (film).
Nollman, Jim. 1999. *The Charged Border: Where Whales and Humans Meet*. New York: Henry Holt and Company.
– 2002. *The Man Who Talks to Whales*. Boulder, CO: Sentient Publications.
Norris, Christopher (ed.) 1989. *Music and the Politics of Culture*: London: Lawrence and Wishart.
– 1997. *Against Relativism: Philosophy of Science, Deconstruction, and Critical Theory*. Malden, Massachusetts: Blackwell.
Novak, Barbara. 1995. *Nature and Culture: American Landscape and Painting, 1825-1875*. New York: Oxford University Press.
Nye, David E. 1994. *American Technological Sublime*. Cambridge, Massachusetts and London, England: MIT Press.
O'Leary, Stephen D. 1994. *Arguing the Apocalypse: A Theory of Millennial Rhetoric*. New York and Oxford: Oxford University Press.
Oakley, Giles. 1983. *The Devil's Music: a History of the Blues*. London: Ariel Books, British Broadcasting Corporation.
Osgerby, Bill. 1999. "Chewing Out A Rhythm on My Bubble-Gum": the Teenage Aesthetic and Genealogies of American Punk' in Sabin (ed.) (1999): 154-69.
Opie, John. 1994. *The Law of the Land: Two Hundred Years of American Farmland Policy*. Lincoln and London: University of Nebraska Press.
Orman, John. 1984. *The Politics of Rock*. Chicago: Nelson-Hall.

Ortiz, John M. 1997. *Tao of Music: Using Music to Change Your Life*. Dublin: Newleaf.

Ovid. 1955. *Metamorphoses* (tr. M. M. Innes). Harmondsworth: Penguin.

Pattison, Robert. 1987. *The Triumph of Vulgarity: Rock Music in the Mirror of Romanticism*. New York and Oxford: Oxford University Press.

Pepper, David. 1996. *Modern Environmentalism: An Introduction*. London and New York: Routledge.

Peretz, Isabel. 2001. 'Listen to the Brain: A Biological Perspective on Musical Emotions' in Juslin and Sloboda (eds). (2001): 105-34.

Peterson, Richard A. 1997. *Creating Country Music: Fabricating Authenticity*. Chicago and London: University of Chicago Press.

Phillips, Dana. 2003. *The Truth of Ecology: Nature, Culture, and Literature in America*. Oxford: Oxford University Press.

Plato. 1974. *The Republic* (tr. D. Lee). Harmondsworth: Penguin.

Plumwood, Val. 1993. *Feminism and the Mastery of Nature*. London and New York: Routledge.

Pouncey, Edwin. 1998. 'Blood on the Frets'. *The Wire*. 174 (August 1998): 22-31.

Pynchon, Thomas. 1973. *Gravity's Rainbow*. London: Picador.

Quinn, Eithne. 2005. *Nuthin' But a 'G' Thang: The Culture and Commerce of Gangsta Rap*. New York: Columbia University Press.

Raitt, Bonnie. (nd). 'Activism'. On-line at: http://www/bonnie raitt.com (consulted 30.04.2009).

Regenstein, Lewis. 1975. *The Politics of Extinction: The Shocking Story of the World's Endangered Wildlife*. New York: Macmillan.

Reich, Charles A. 1971. *The Greening of America*. London: Allen Lane.

Reuss, Richard A. with JoAnne C. Reuss. 2000. *American Folk Music and Left Politics 1927-1957*. Lanham, Maryland and London: Scarecrow Press.

Reynolds, Simon. 1999. *Generation Ecstasy: Into the World of Techno and Rave Culture*. New York: Routledge.

Reynolds, Simon and Joy Press. 1995. *The Sex Revolts: Gender, Rebellion and Rock 'n' Roll*. London: Serpent's Tail.

Rimbaud, Arthur. 2005. *Selected Poems and Letters* (tr. Sturrock J. and Harding, J.). Harmondsworth: Penguin.

Robertson, George, Melinda Mash, Lisa Tickner, Jon Bird, Barry Curtis and Tim Putnam (eds). 1996. *FutureNatural: Nature, Science, Culture*. London and New York: Routledge.

Robertson, Roland. 1992. *Globalization: Social Theory and Global Culture*. London: Sage.

Rodaway, Paul. 1994. *Sensuous Geographies: Body, Sense and Place*. London and New York: Routledge.

Rose, Tricia. 1994. *Rap Music and Black Culture in Contemporary America*. Hanover and London: Wesleyan University Press.

Ross, Andrew. 1991. *Strange Weather: Culture, Science and Technology in the Age of Limits*. London and New York: Verso.

Rossinow, Doug. 1998. *The Politics of Authenticity: Liberalism, Christianity, and the New Left in America*. New York: Columbia University Press.

Roszak, Theodore. 1970. *The Making of a Counter Culture: Reflections on the Technocratic Society and Its Youthful Opposition*. London: Faber and Faber.

– 2001. *The Voice of the Earth: an Exploration of Ecopsychology*. Grand Rapids, Michigan: Phanes.

Rothenberg, David. 1993. *Hand's End: Technology and the Limits of Nature*. Berkeley, Los Angeles, London: University of California Press.

– 2000. "No World But in Things": The Poetry of Naess's Concrete Contents' in Katz, Eric, Andrew Light and David Rothenberg (eds). *Beneath the Surface: Critical Essays in the Philosophy of Deep Ecology*. Cambridge and London: MIT Press: 151-68.

– 2001. 'Introduction: Does Nature Understand Music?' in Rothenberg, David and Marta Ulvaeus (eds). *The Book of Music and Nature*. Middletown, CT: Wesleyan Press: 1-10.

– 2002. *Sudden Music: Improvisation, Sound, Nature*. Athens: University of Georgia Press.

– 2005. *Why Birds Sing: One Man's Quest to Solve an Everyday Mystery*. Harmondsworth: Penguin.

– 2008. *Thousand Mile Song: Whale Music in a Sea of Sound*. New York: Basic Books.

Rouget, Gilbert. 1985. *Music and Trance: A Theory of the Relations between Music and Possession* (tr. B. Biebuyck). Chicago and London: Chicago University Press.

Rucker, Rudy, R. U. Sirius and Queen Mu. 1992. *Mondo 2000: A User's Guide to the New Edge*. New York: HarperPerennial.

Rudland, Dean. 2001. 'David Axelrod'. Liner notes to *Earth Rot*. Stateside. 2001. (CD).

Rushkoff, Douglas. 1994. *Cyberia: Life in the Trenches of Hyperspace*. London: Flamingo.

Russolo, Luigi. 1986. *The Art of Noises* (tr. B. Brown). New York: Pendragon Press.

Sabin, Roger (ed.) 1999. *Punk Rock: So What?* London and New York: Routledge.

Said, Edward W. 1985. *Orientalism*. Harmondsworth: Penguin.

Schwarz, David. 1997. *Listening Subjects: Music, Psychoanalysis, Culture*. Durham and London: Duke University Press.

Schopenhauer, Arthur. 1969. *The World as Will and Representation*. Vol. 1. (tr. E. P. J. Payne). New York: Dover Publications.

Seeger, Pete (as Johnny Appleseed, Jr.). 'Found: A Prolific New Songwriter'. *Sing Out!* 9.2 (Fall 1959): 41.

– 1963a. *Sing Out!* 13.3 (Summer 1963): 67.

– 1963b. *Sing Out!* 13.4 (October 1963): 65.

– 1963c. *Sing Out!* 16.6 (December 1967): 54.

– 1969. 'My Hudson River (My Dirty Stream)'. *Sing Out!* 19.2 (July 1969): 1.

– 1993. *Where Have All The Flowers Gone?: A Singer's Stories, Songs, Seeds, Robberies*. Bethlehem, PA: Sing Out Corporation.

Shapiro, Peter. 2005. *The Rough Guide to Hip-Hop*. London: Rough Guides.

Shepherd, John and Peter Wicke. 1997. *Music and Cultural Theory*. Cambridge: Polity Press.

Silber, Irwin. 1968. 'The Topical Song Revolution and How it Fizzled Out' in *Guardian* (September 28 1968): 20-1.

Silverman, Kaja. 1988. *The Acoustic Mirror: the Female Voice in Psychoanalysis and Cinema*. Bloomington: Indiana University Press.

Sinclair, John. 1972. *Guitar Army: Street Writings / Prison Writings*. New York: Douglas Book Corp.

Sloboda, John A. 1985. *The Musical Mind: The Cognitive Psychology of Music*. Oxford: Clarendon Press.
Smith, Christine. 2001. *A Mountain in the Wind: An Exploration of the Spirituality of John Denver*. Forres: Findhorn Press.
Smith, Neil. 1996. 'The Production of Nature' in Robertson, George, Melinda Mash, Lisa Tickner, Jon Bird, Barry Curtis and Tim Putnam (eds). (1996): 35-54.
Smucker, Tom. 1969. 'Bob Dylan Meets the Revolution' in McGregor, Craig (ed.) *Bob Dylan: A Retrospective* (1980). London: Angus & Robertson: 179-183.
Sontag, Susan. 1967. *Against Interpretation*. London: Eyre and Spottiswoode.
Soper, Kate. 1996. 'Nature / "nature" in Robertson, George, Melinda Mash, Lisa Tickner, Jon Bird, Barry Curtis and Tim Putnam (eds). (1996): 22-34.
– 1995. *What is Nature? Culture, Politics and the Non-Human*. Oxford, UK and Cambridge, Massachusetts: Blackwell.
St. John, Graham. 2001. *Free NRG: Notes from the Edge of the Dance Floor*. Altona: Common Ground.
Stephens, Julie. 1998. *Anti-Disciplinary Protest: Sixties Radicalism and Postmodernism*. Cambridge: Cambridge University Press.
Stoett, Peter J. 1994. 'Cities: To Love or to Loathe?' *Environmental Politics*. 3 (2): 339-42.
– 1995. *Atoms, Whales, and Rivers: Global Environmental Security and International Organization*. New York: Nova.
Stokes, Martin. 2003. 'Globalization and the Politics of World Music' in Clayton, Martin, Trevor Herbert and Richard Middleton (eds). *The Cultural Study of Music: A Critical Introduction*. New York and London: Routledge.
Szwed, John F. 2000. *Space is the Place: the Lives and Times of Sun Ra*. Edinburgh: Mojo Books.
Tagg, Philip. 1982. 'Nature as a Musical Mood Category'. On-line at: http://www.tagg.org/articles/nature.html (consulted 05.03.2009).
– 1989. 'Open Letter about 'Black Music', 'Afro-American Music' and 'European Music'. On-line at: http://www.tagg.org/articles/opelet.html (consulted 05.03.2009).
Tamarkin, Jeff. 2003. *Got a Revolution! The Turbulent Flight of Jefferson Airplane*. London: Helter Skelter.
Taylor, Arthur. 1982. *Notes and Tones: Musician-to-Musician Interviews*. London, Melbourne and New York: Quartet Books.
Taylor, Timothy D. 1997. *Global Pop: World Music, World Markets*. New York and London: Routledge.
Thompson, William Irwin. 1989. 'Pacific Shift' in Callicott, J. Baird and Roger T. Ames (eds). (1989): 25-36.
Tichi, Cecelia. 1994. *High Lonesome: The American Culture of Country Music*. Chapel Hill and London: University of North Carolina Press.
Tompkins, Peter and Christopher Bird. 1973. *The Secret Life of Plants*. London: Allen Lane.
Toop, David. 1995. *Ocean Of Sound: Aether Talk, Ambient Sound and Imaginary Worlds*. London: Serpent's Tail.
– 2004. *Haunted Weather: Music, Silence and Memory*. London: Serpent's Tail.
Torgovnick, Marianna. 1990. *Gone Primitive: Savage Intellects, Modern Lives*. Chicago and London: University of Chicago Press.

Tosches, Nick. 1985. *Country: Living Legends and Dying Metaphors in America's Biggest Music*. London: Secker and Warburg.

Toynbee, Jason. 2000. *Making Popular Music: Musicians, Creativity and Institutions*. London: Arnold.

Trakin, Roy. 1983. 'The Teachings of Captain Beefheart: A Wacky Way of Knowledge'. On line at: http://www.beefheart.com (consulted 06.08.06).

*Triumph of the Will* 1935. Dir. Leni Riefenstahl. NSDAP. Germany. (film).

Troy, Sandy. 1994. *Captain Trips: The Life and Fast Times of Jerry Garcia*. London: Virgin Books.

Vale, V. and Andrea Juno (eds). 1994. *RE/SEARCH Incredibly Strange Music Volume II*.San Francisco: RE/SEARCH Publications.

Wallin, Nils, Björn Merker and Steven Brown (eds). 2000. *The Origins of Music*. Cambridge, Massachusetts and London: MIT Press.

Ward, Brian. 1998. *Just My Soul Responding: Rhythm and Blues, Black Consciousness and Race Relations*. London: UCL Press.

Ward, Geoff. 1993. *Language Poetry and the American Avant-garde*. Keele: British Association for American Studies.

Watson, Ben. 1994. *Frank Zappa: The Negative Dialectics of Poodle Play*. London: Quartet Books.

Webb, Jimmy. 1998. *Tunesmith: Inside the Art of Songwriting*. New York: Hyperion.

Webster, Duncan. 1988. *Looka Yonder! The Imaginary America of Populist Culture*. London and New York: Routledge.

Whitley, David. 2008. *The Idea of Nature in Disney Animation*. Aldershot: Ashgate.

Wilkie, Richard W. 1969. 'The Hudson River Sloop Restoration'. *Sing Out!* 19.2. (July 1969): 2-3.

Willener, Alfred. 1970. *The Action-Image of Society: On Cultural Politicization*. London: Tailstock Publications.

Williams, Raymond. 1983. *Keywords: A Vocabulary of Culture and Society*. London: Fontana Press.

Williams, William Carlos. 1954. *Selected Essays*. New York: Random House.

Willis, Paul. 1978. *Profane Culture*. London: Routledge and Kegan Paul.

Wilson, Edward O. 1984. *Biophilia*. Cambridge, Massachusetts and London: Harvard University Press.

Winner, Langdon. 1970a. Review of *Déjà Vu*. *Rolling Stone* (April 30 1970): 53.

– 1970b. 'I'm not even here I just stick around for my friends: The Odyssey of Captain Beefheart'. *Rolling Stone* (May 14 1970). On-line at: http://www.beefheart.com (consulted 11.06.2009).

– 1971. 'In Search of America: Captain Beefheart and the Smithsonian Institute Blues'. *Rolling Stone* (April 1 1971). On-line at: http://www.members.lycos.nl/teejo/argue/search.html (consulted 11.06.2009).

Wolfe, Tom. 1969. *The Electric Kool-Aid Acid Test*. New York: Bantam.

Woods, Clyde. 1998. *Development Arrested: The Blues and Plantation Power in the Mississippi Delta*. London and New York: Verso.

– 2007. 'Sittin' on Top of the World: The Challenges of Blues and Hip Hop Geography' in McKittrick, Katherine and Clyde Woods (eds). *Black Geographies and the Politics of Place*. Cambridge, Massachusetts: South End Press: 46-81.

Zappa, Frank with Peter Occhiogrosso. 1989. *The Real Frank Zappa Book*. London: Picador.

Zuckerkandl, Victor. 1969. *Sound and Symbol: Music and the External World.* Princeton: Princeton University Press.
– 1973. *Man the Musician*. Princeton: Princeton University Press.

# Index

CPSIA information can be obtained at www.ICGtesting.com
Printed in the USA
LVOW07s2303091214

418087LV00002B/257/P

9 789042 032095